Boxes, Baskets, Pots & Planters

A practical guide to 100 inspirational containers

STEPHANIE DONALDSON

PREMIER

This edition published in the UK in 1997 by Hermes House

Hermes House is an imprint of
Anness Publishing Limited
Hermes House
88-89 Blackfriars Road
London SE1 8HA

ISBN 1 901289 12 5

A CIP catalogue record for this book is available from the British Library

Publisher: Joanna Lorenz
Editorial Manager: Helen Sudell
Designers: Blackjacks, Lillian Lindblom, Peter Laws
Photographers: Marie O'Hara and John Freeman (finished shots),
Janine Hosegood (step photography)
Stylist: Stephanie Donaldson

Printed in Hong Kong

1 3 5 7 9 10 8 6 4 2

Boxes, Baskets, Pots & Planters

Contents

Introduction

You don't need expert knowledge to be a successful container gardener. Indeed, if you own a houseplant or pot of parsley, you are already container gardening. Today, good quality container-grown plants are available all through the year, and skills such as seed sowing, potting-on and taking cuttings, although interesting, are no longer essential. There is nothing wrong with this form of "instant" gardening. By buying plants that are nearing maturity, you are saved the time-consuming and uncertain business of rearing the plants, and you can try out different shape and colour combinations before you buy the plants by simply grouping them together.

Gardening in hanging baskets, pots and planters has other advantages. Like furniture, containers can be moved around to create a new look, with pride of place given to those plants that are performing best. Unlike plants in the border, the containers can simply be moved into the background once flowering is over.

The plants are only a part of the equation. You can have fun being creative with your choice of containers. Shops and garden centres stock an ever-increasing range of containers to suit every budget, from magnificent Italian olive jars to humble plastic pots. Junk shops and garage sales are also a rich source of objects that can take on a new lease of life as a planter – a blackened cooking pot with a hole in its base is useless for its original purpose but is ideal as a container of character and can be bought cheaply.

Whether you are a novice or an experienced gardener, container gardening is an accessible, enjoyable and colourful way to brighten up your surroundings.

Seed Sowing

Some plants are very easy to sow from seed – marigolds rarely disappoint, even if you are a complete beginner.

1 Fill the pot with seed compost. Gently firm and level the surface by pressing down on the compost using a pot of the same size.

2 When sowing large seeds, such as sunflowers or marigolds, use a dibber, cane or pencil to make holes for each seed. Plant the seeds and then firmly tap the side of the pot with the flat of your hand to fill the holes with compost. Water from above, using a fine rose on a watering can, or by standing the pot in a saucer of water until the surface of the compost is moist. Cover the pot with a black plastic bag as most seeds germinate best in a warm dark place. Check daily and bring into the light when the seedlings are showing.

3 When sowing small seeds they should be thinly scattered on the surface of the compost and then covered with just enough sieved compost to conceal them. Firm the surface, using another pot, and then treat in the same way as large seeds.

Potting-on

Sooner or later, plants need repotting. Young seedlings, shown here, do not thrive in over-large pots. Divide the plants, if necessary, and plant them in pots the same size as the one they were previously grown in.

1 Seedlings will probably be ready to move into larger pots when the roots start to emerge through the holes in the base of the pot. To check, gently remove the rootball from the pot and, if there are plenty of roots showing, you will know the plants are ready for a move.

2 If there is more than one seedling in the pot, gently break each seedling away with a good rootball. (Some plants hate to have their roots disturbed. The information on the seed packet will tell you this. These seeds are best sown individually in peat pots.)

3 Lower the rootball of the plant into the pot and gently pour compost around it, lightly pressing the compost around the roots and stem. It does not matter if the stem of the seedling is buried deeper than it was previously as long as the leaves are well clear of the soil. Water, using a can with a fine rose.

Watering

Like other containers, window boxes dry out very quickly and regular watering is essential. Watering should be carried out in the early morning or late evening during summer months. If only one watering is possible, an evening watering is preferable as the plants have the cool night hours in which to absorb the water.

1 Watering a large window box with a hose is easier and more effective than a watering can, provided there are no restrictions on hose use.

2 Small window boxes can be adequately watered with a watering can.

Insecticides

There are two main types of insecticide available to combat common pests. Before planting your window box it is advisable to check the plants for pests and, if any are found, follow the recommended treatment. During the growing season, keep a look-out for pests and treat your plants before any real damage is done.

Contact insecticides

These must be sprayed directly on to the insects to be effective. Most organic insecticides work this way, but they generally kill all insects, even beneficial ones, such as hoverflies and ladybirds. Try to remove these before spraying the infected plant.

Systemic insecticides

These work by being absorbed by the plant's root or leaf system and killing the insects that come into contact with the plant. This will work for difficult pests, such as vine weevils which are hidden in the soil, and scale insects which protect themselves from above with a scaly cover.

BIOLOGICAL CONTROL

Commercial growers now use biological control in their glasshouses; this involves natural predators being introduced to eat the pest population. Although not all of these are suitable for the amateur gardener, they can be used in conservatories for dealing with pests such as whitefly.

Composts

Composts come in various formulations suitable for different plant requirements. A standard potting compost is usually peat-based and is suitable for all purposes. Different composts can be mixed together for specific plant needs.

Standard compost
The majority of composts available at garden centres are peat-based with added fertilizers.

Ericaceous compost
This is a peat-based compost with no added lime, essential for rhododendrons, camellias and heathers in containers.

Container compost
This is a peat-based compost with moisture-retaining granules and added fertilizer, specially formulated for window boxes and containers.

Loam-based compost
This uses sterilized loam as the main ingredient, with fertilizers to supplement the nutrients in the loam. Although much heavier than peat-based compost, it can be lightened by mixing with peat-free compost. Ideal for long-term planting.

Peat-free compost
Manufacturers are beginning to offer composts using materials from renewable resources such as coir fibre. They are used in the same way as peat-based composts.

Mulches

A mulch is a layer of protective material placed over the soil. It helps to retain moisture, conserve warmth, suppress weeds and prevent soil splash on foliage and flowers.

Composted bark
Bark is an extremely effective mulch and, as it rots down, it also conditions the soil. It works best when spread at least 8 cm (3 in) thick and is therefore not ideal for small containers. It is derived from renewable resources.

Clay granules
Clay granules are widely used for hydroculture, but can also be used to mulch house plants. When placing a plant in a cachepot, fill all around the pot with granules. When watered, the granules absorb moisture, which is then released slowly to create a moist microclimate for the plant.

Stones
Smooth stones can be used as decorative mulch for large container-grown plants. You can save stones dug out of the garden or buy stones from garden centres. Cat owners will also find they keep cats from using the soil surrounding large house plants as a litter tray.

Gravel
Gravel makes a decorative mulch for container plants and provides the correct environment for plants such as alpines. It is available in a variety of sizes and colours which can be matched to the scale and colours of the plants used.

Water-retaining Gel

One of the main problems with window boxes is the amount of watering required to keep the plants thriving. Adding water-retaining gels to the compost will certainly help to reduce this task. Sachets of gel are available from garden centres.

1 Pour the recommended amount of water into a bowl.

2 Scatter the gel over the surface, stirring occasionally until it has absorbed the water.

3 Add to your compost at the recommended rate.

4 Mix the gel in thoroughly before using it for planting.

Saucers and Feet

Saucers are available for plastic and clay pots. They act as water reservoirs for the plants, and are used under houseplants to protect the surface they are standing on. Clay saucers must be fully glazed if they are used indoors or they will leave marks. Clay feet are available for terracotta pots. They will prevent the pot becoming waterlogged, but this also means that in a sunny position the pot will dry out very quickly and may need extra watering.

1 Plastic saucers can be used to line containers which are not waterproof, such as this wooden apple basket.

Common Pests

Caterpillars (*above*)

The occasional caterpillar can simply be picked off the plant and disposed of as you see fit, but a major infestation can strip a plant before your eyes. Contact insecticides are usually very effective in these cases.

Snails (*right*)

Snails can be a problem in window boxes as they tuck themselves behind the containers during daylight and venture out to feast at night. Slug pellets should deal with them or, alternatively, you can venture out yourself with a torch and catch them.

Vine weevils (*above*)

These white grubs are a real problem. The first sign of an infestation is the sudden collapse of the plant, which has died as a result of the weevil eating its roots. Systemic insecticides or natural predators can be used as a preventative, but once a plant has been attacked it is usually too late to save it. Never re-use the soil from an affected plant.

Whitefly

As their name indicates, these are tiny white flies which flutter up in clouds when disturbed from their feeding places on the underside of leaves. Whitefly are particularly troublesome in conservatories, where a dry atmosphere encourages them to breed. Keep the air as moist as possible. Contact insecticides will need more than one application to deal with an infestation, but a systemic insecticide will protect the plant for weeks.

Mealy bugs

These look like spots of white mould. They are hard to shift and regular treatment with a systemic insecticide is the best solution.

Greenfly

One of the most common plant pests, these green sap-sucking insects feed on the tender growing tips of plants. Most insecticides are effective against greenfly. Choose one that will not harm ladybirds as greenfly are a favourite food of theirs.

Red spider mite (*above*)

This is another insect that thrives indoors in dry conditions. Constant humidity will reduce the chance of an infestation. The spider mite is barely visible to the human eye, but infestation is indicated by the presence of fine webs and mottling of the plant's leaves. To treat an infestation, pick off the worst affected leaves and spray the plants with an insecticide.

Feeding your Plants

It is not generally known that most potting composts only contain sufficient food for six weeks of plant growth. After that, the plants will slowly starve unless other food is introduced. There are several products available, all of which are easy to use. All the projects in this book use slow-release plant food granules as this is the easiest and most reliable way of ensuring your plants receive sufficient food during the growing season. For these granules to be effective the compost needs to remain damp or the nutrients cannot be released.

A variety of plant foods: from the left; liquid feed, 2 types of pelleted slow-release plant food granules, a general fertilizer and loose slow-release plant food granules.

Slow-release plant food granules
These will keep your window box plants in prime condition and are very easy to use. One application lasts six months, whereas most other plant foods need to be applied fortnightly. Follow the manufacturer's recommended dose carefully; additional fertilizer will simply leach away.

When adding fertilizer granules to the soil, sprinkle them on to the surface of the compost and rake into the top layer. Pelleted granules should be pushed approximately 2 cm (1 in) below the surface.

Liquid feeds
These are available in many formulations. Generally, the organic liquid manures and seaweed feeds are brown in colour and should be mixed to look like very weak tea. The chemical feeds are frequently coloured to prevent them being mistaken for soft drinks. The best way to avoid accidents with garden chemicals is to mix up only as much as you need on each occasion and never store them in soft-drinks bottles. Liquid feeds should be applied fortnightly in the growing season. Do not mix a feed stronger than is recommended – it can burn the roots of the plant and it certainly will not make it grow any faster. These can be used in addition to granules for really lush results.

Planting in Terracotta

Terracotta window boxes are always popular, but need some preparation before planting.

1 With terracotta window boxes it is essential to provide some form of drainage material in the base. In small window boxes this can be broken pieces of pot, known as crocks.

2 Another useful drainage material is gravel, which is easily available from your local garden centre in various sizes.

3 When planting in large window boxes, recycle polystyrene plant trays as drainage material. Lumps of polystyrene are excellent for this purpose and as they retain warmth they are of additional benefit to the plant.

4 Cover the drainage material with a layer of compost before planting in the window box.

Planting Wicker Baskets

If you wish to use a more unconventional container as a window box you may need to seal it in some way to prevent leakage.

1 Line the basket with a generous layer of moss – this will prevent the compost leaking away.

2 Fill the basket with compost and mix in plant food granules or any organic alternative you wish to use.

Planting in Plastic Window Boxes

If you do not wish to use heavy terracotta containers, there are plenty of good-sized plastic ones available. It is also cheaper to use one of these and then place it inside a more attractive, but possibly old or fragile, container.

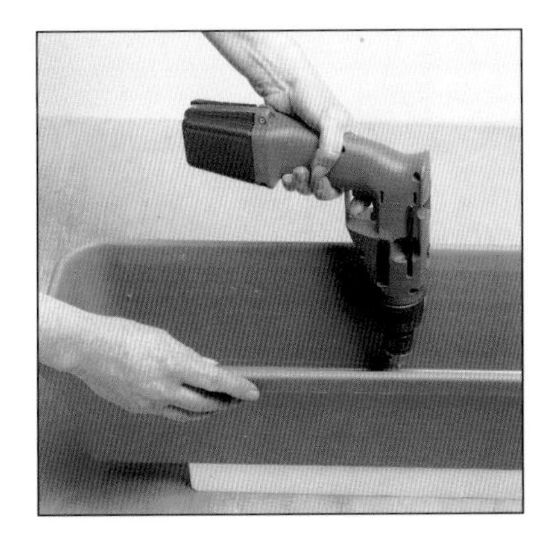

1 When buying plastic pots, check that the drainage holes are open. Some manufacturers mark the holes, but leave it to the customer to punch them out or drill them as required. If using a drill, remember to place a good wedge of polystyrene or some other material between the pot and the top of your work table.

2 With plastic window boxes there is no need to use any drainage material at the base of the container; simply cover the bottom of the pot with a layer of compost.

Plant Supports

Climbing plants need support even in window boxes. Support can be provided by using canes, which can be pushed into the window box, or a trellis, which is fastened to a wall, or a free-standing frame.

Plants that are Pot-bound

Some plants that have been growing in small pots for a certain length of time can become "pot-bound". Gently tease out the roots around the bottom and edges to encourage the roots to grow down into the container.

Preparing the Basket

The key to successful hanging baskets is in the preparation. Time taken in preparing the basket for planting will be rewarded with a long-lasting colourful display. Slow-release plant food granules incorporated into the compost when planting will ensure that the plants receive adequate nutrients throughout the growing season. It is essential to water hanging baskets every day, even in overcast weather, as they dry out very quickly. There are various ways to line a hanging basket, but the most attractive and successful is to use moss. All of the baskets in this book are lined with sphagnum moss - available from most garden centres and good florists.

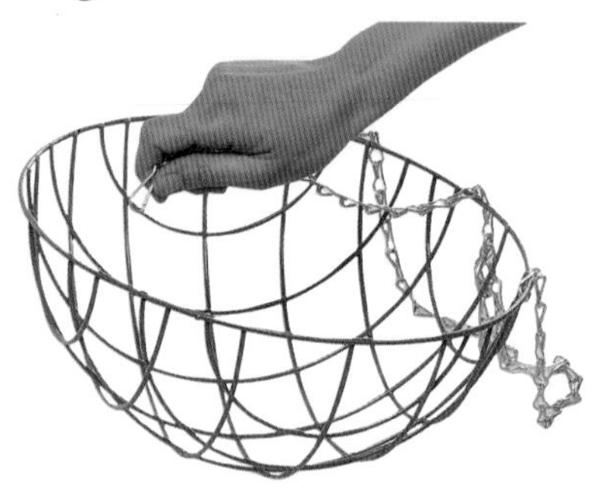

When buying a hanging basket, make sure that the chains are detachable. Unhooking one of the chains enables them to be placed to one side of the basket, allowing you to work freely.

Alternative Linings

If you cannot find sphagnum moss, or if you want to use something else to line your baskets, there are other options.

1 Although not as good to look at as moss, coir fibre lining is soon hidden as the plants grow. The slits allow for planting in the side of the basket.

2 Cardboard liners are clean and easy to use. They are made in various sizes to fit most hanging baskets.

3 Press out the marked circles on the cardboard liner if you wish to plant into the side of it.

Underplanting a Hanging Basket

Underplanting helps to achieve a really lush-looking basket and soon conceals the shape of the container under flowers and foliage.

1 Line the lower half of the basket with a generous layer of moss.

2 Gently guide the foliage through the sides of the basket and rest the rootball on the moss.

3 Add more moss to the basket, tucking it carefully around the plants to ensure that they are firmly in place. Add a further row of plants near the top edge of the basket, if required, and continue to line the basket with moss, finishing off with a collar of moss overlapping the rim of the basket. Fill with compost.

Planting a Wall Basket

The principle of planting a wall basket is the same as an ordinary one, but for maximum effect it is always a good idea to underplant as well to give a good display of colour against the wall.

1 Line the back and lower half of the front of the basket with moss.

2 Plant some of your chosen plants into the side of the basket by resting the rootballs on the moss and feeding the foliage through. Add another layer of soil and moss and plant more plants.

3 Complete lining the basket with moss and fill with compost mixture and any slow-release feed you may wish to use. Plant the rest of your plants, starting at the centre and working out to the sides.

Planting a Hanging Basket

1 Tease out the moss and fill the bowl of the basket with a generous layer. It is important to make sure there are no holes or compost and water will escape.

2 Build a thick collar of moss overlapping the rim of the basket. This will ensure that water soaks into the basket rather than running off the surface.

3 Fill the basket with compost.

Types of Container

Part of the fun of container gardening is experimenting with the different planters available. Garden centres stock an increasing variety of styles, and junk shops, car boot sales and flea markets are also worth a visit.

Novelty containers
Old gardening or cooking implements which are no longer suitable for their original purpose can be spray-painted or stencilled.

Terracotta pots
Almost every style and size imaginable is available in terracotta, from huge floor-standing planters for trees, to simple rustic flower pots and wall planters.

Painted clay pots
These will add an additional spot of colour in the house or garden. Buy ready-painted or paint your own in soft pastels or bright Mediterranean colours.

Baskets
These add a wonderful country feel to any display, and look delightful planted with spring bulbs. Always line before planting, or use simply as a cachepot.

Wire baskets
Lined with sphagnum moss, these make pretty containers for spring bulbs or other small flowering plants.

Galvanized tinware
This is available both new and secondhand, in all shapes and sizes, from small buckets to tin baths; small planters to free standing pots

Window boxes
These are available in a range of different materials, from terracotta to bark. This planter has a verdigris effect, but is in fact made of lightweight fibreglass.

Types of Window Boxes

Think of the setting when you buy a window box - a rustic wooden planter will look wonderful under the window of a thatched cottage but may look very out of place in front of an elegant town house. Bear proportions in mind, too, and look for a window box that fits comfortably on your sill or bracket without looking crowded or lost in its surroundings.

Wooden window boxes
These are not as popular as they used to be. Styles vary from rustic to very sophisticated.
Advantages - ages attractively providing it has been made with treated wood, relatively lightweight.
Disadvantages - some maintenance necessary.

Terracotta window boxes
These are available in a wide range of sizes and styles.
Advantages - looks good and will look even better with age.
Disadvantages - heavy and may be damaged by frost.

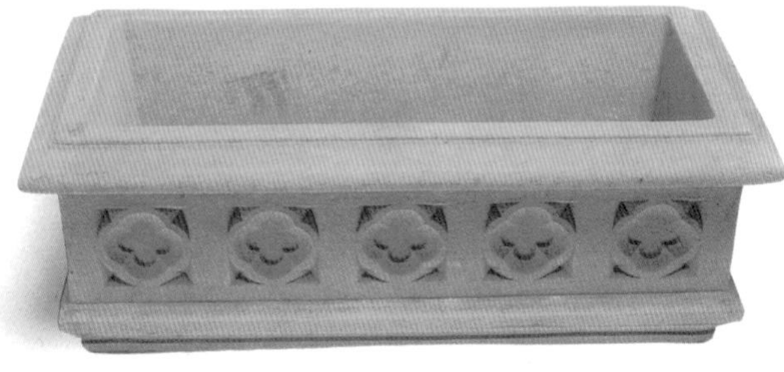

Stone window boxes
Not so readily available but worth looking out for.
Advantages - durable and attractive.
Disadvantages - very heavy and expensive.

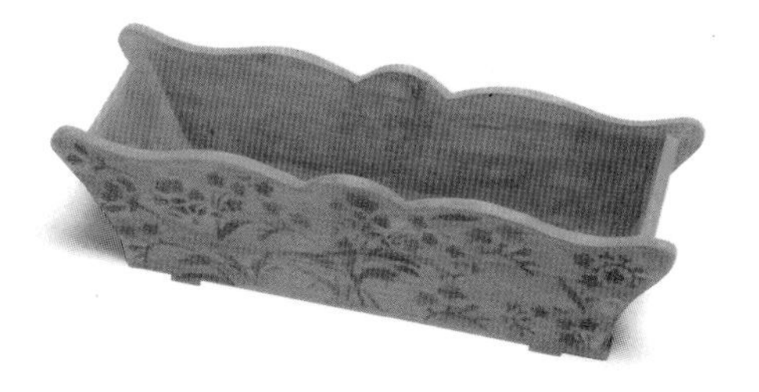

Painted wooden window boxes
These can be customized by painting to make an original and unusual container.
Advantages - your window box will be a one-off, and you can change its look next year; relatively lightweight.
Disadvantages - will need suitable paint and materials and window box will require some ongoing maintenance.

Bark window boxes
These are stylish and wonderfully natural looking.

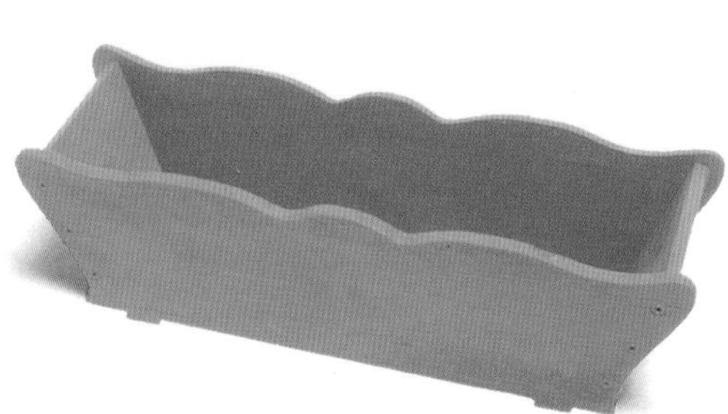

Lightweight fibre window boxes
These are a practical alternative to plastic, although they will not last as long.

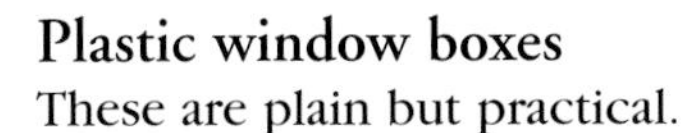

Plastic window boxes
These are plain but practical.
Advantages - maintenance free and very lightweight.
Disadvantages - not very exciting and will not age attractively.

Baskets
These can be used as window boxes provided they are generously lined with moss before planting.

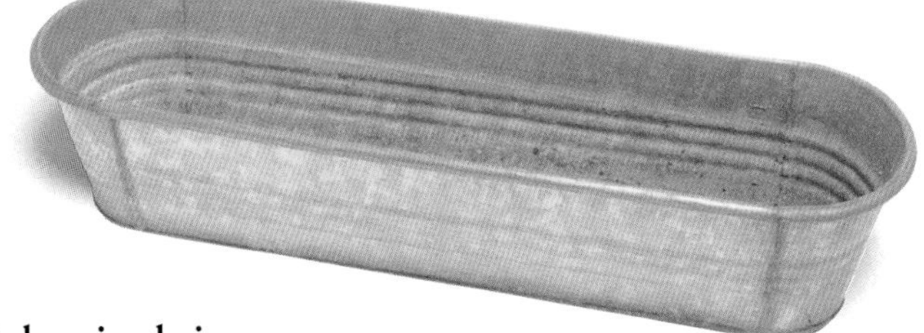

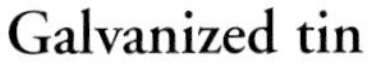

Galvanized tin
Tin has moved from the utilitarian to the fashionable and a tin window box is an interesting variation from the usual materials.

Types of Hanging Basket

Before you choose the plants and how to arrange them, you need to decide what sort of hanging basket you are going to display them in. Garden centres stock a huge variety, which are all easy to work with and hang.

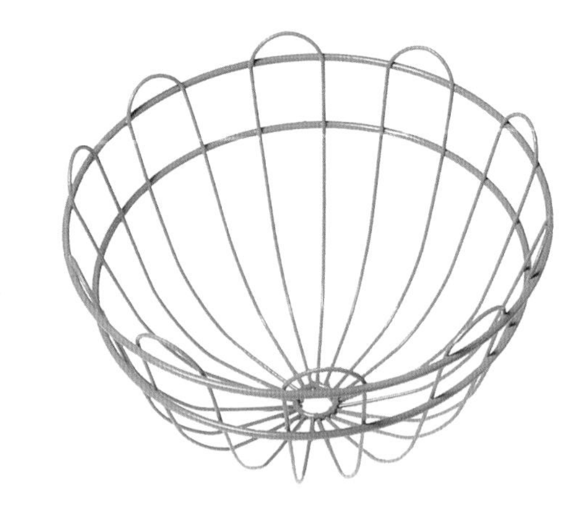

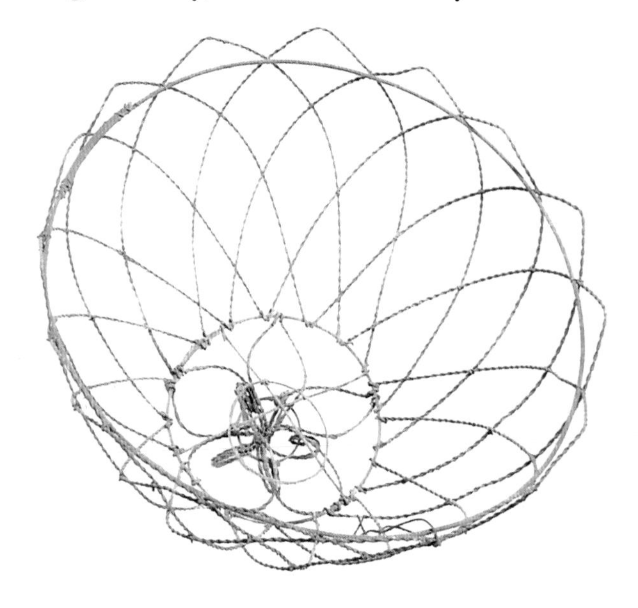

Baskets are generally made from plastic-coated wire, but are also available in wrought iron and galvanized wire.

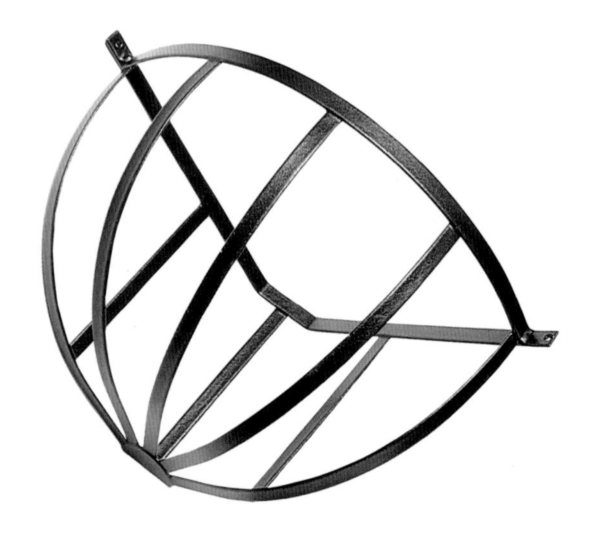

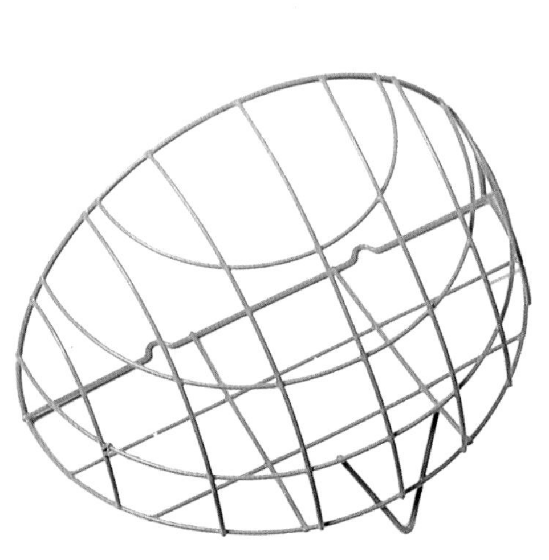

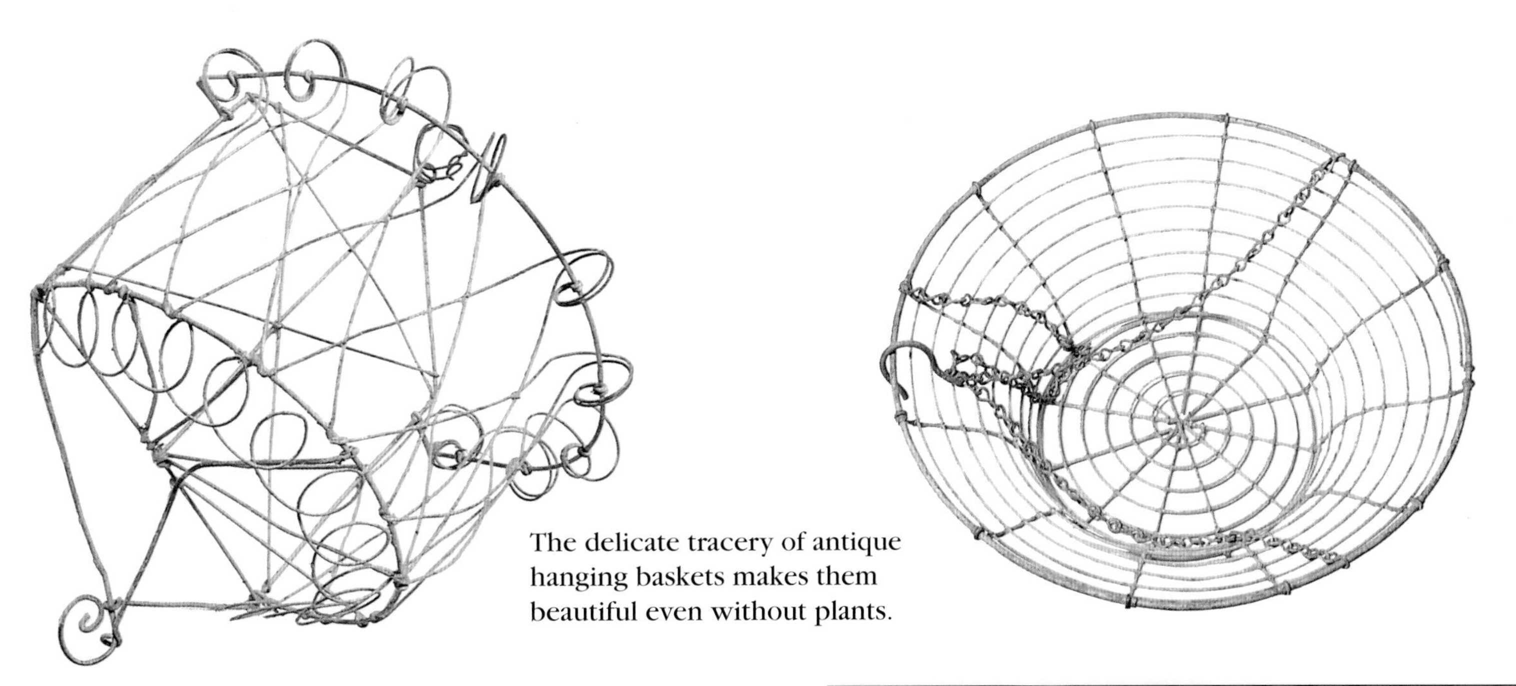

The delicate tracery of antique hanging baskets makes them beautiful even without plants.

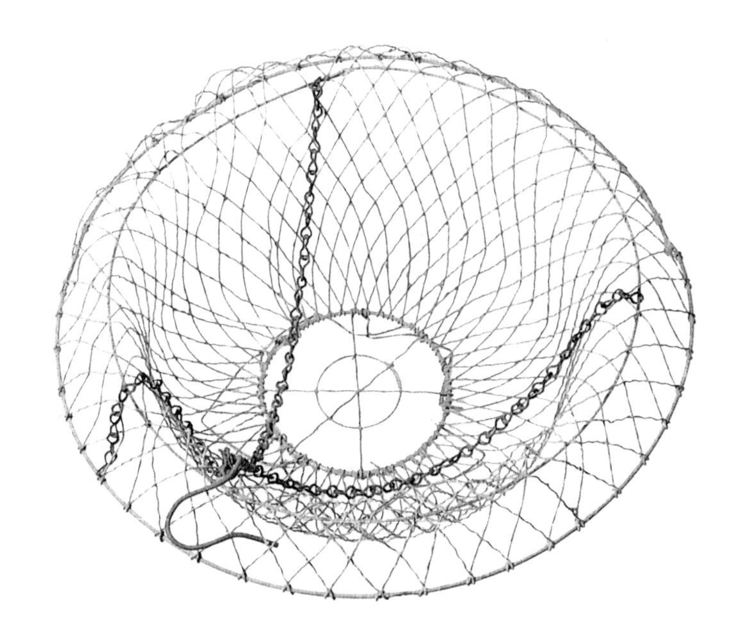

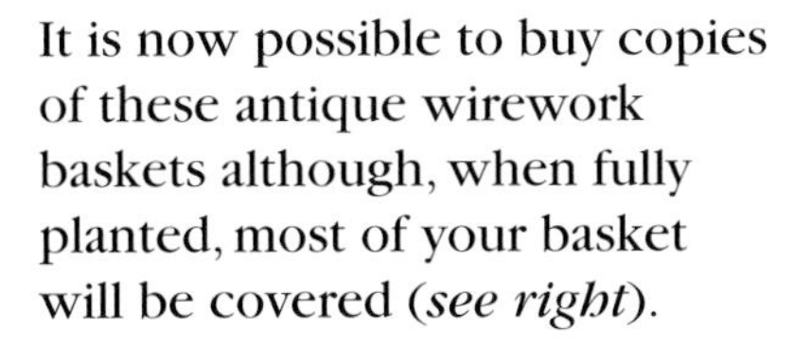

It is now possible to buy copies of these antique wirework baskets although, when fully planted, most of your basket will be covered (*see right*).

A White Theme for an Antique Basket

An antique basket is not essential for this scheme, but the bowl shape makes an interesting variation. A small variety of dahlia, known as dahlietta, has large white flowers that blend with trailing verbenas and a begonia. Silver-leaved *Argyranthemum* brings a subtle touch of colour to the basket.

MATERIALS
36 cm (14 in) hanging basket
Sphagnum moss
Compost
Slow-release plant food granules

PLANTS
White *Begonia semperflorens*
3 white dahlietta
3 white trailing verbena
3 *Argyranthemum* 'Flamingo'

1 Line the basket with moss and fill with compost. Mix a teaspoon of slow-release plant food granules into the top of the compost.

2 Plant the begonia in the centre and position and plant the dahliettas around the begonia.

GARDENER'S TIP

Regular dead-heading of the flowers will keep the basket in tip-top condition. The slow-release plant food granules will give the flowers a regular supply of food provided the basket is not allowed to dry out.

Plant in late spring or early summer

3 Plant the verbenas to one side of each of the dahliettas.

4 Finally plant the *Argyranthemums*, angling the plants to encourage them to trail over the edge of the basket. Water well and hang in a sunny position.

A Silver and White Wall Basket

The silvery *Helichrysum* foliage and cool blue lavender flowers give a delicate colour scheme which would look good against a weathered background.

MATERIALS
30 cm (12 in) wall basket
Sphagnum moss
Compost
Slow-release plant food granules

PLANTS
2 lavender (*Lavandula dentata* var. *candicans*)
Osteospermum 'Whirligig'
2 *Helichrysum petiolare*

Osteospermum

lavender

Helichrysum

GARDENER'S TIP

The lavender used in this project is fairly unusual – if you wish, you can substitute it with a low-growing variety such as 'Hidcote'. Keep the *Helichrysum* in check by pinching out its growing tips fairly regularly or it may take over the basket.

Plant in spring

1 Line the basket with moss.

2 Half fill the basket with the compost. Mix a half-teaspoon of plant food granules into the compost. Plant the lavenders in either corner.

3 Plant the *Osteospermum* in the centre of the basket.

4 Plant the *Helichrysum* on either side of the *Osteospermum* and angle the plants to encourage them to trail over the side of the basket. Water well and hang in a sunny spot.

White Flowers and Painted Terracotta

There are plenty of inexpensive window boxes available, but they do tend to be rather similar. Why not customize a bought window box to give it a touch of individuality? This deep-blue painted window box creates an interesting setting for the cool white geraniums and verbenas.

MATERIALS
45 cm (18 in) terracotta window box painted blue
Crocks or other suitable drainage material
Compost
Slow-release plant food granules

PLANTS
White geranium (*Pelargonium*)
2 variegated *Felicia*
2 white trailing verbena

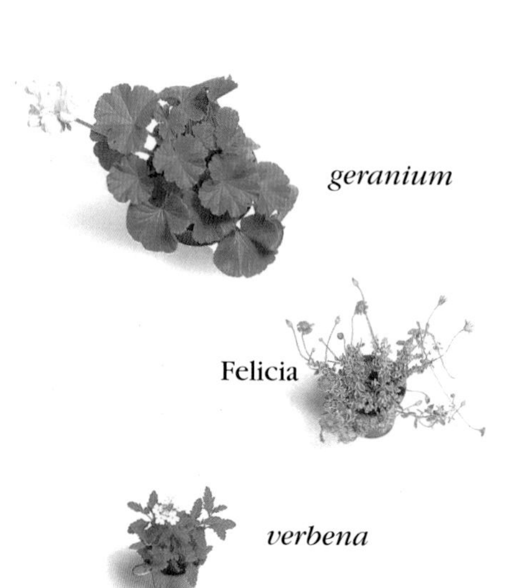

1 Cover the base of the window box with a layer of crocks or similar drainage material.

2 Fill the window box with compost, mixing in 2 teaspoons of slow-release plant food granules. Plant the geranium in the centre of the window box.

3 Plant a *Felicia* on either side of the geranium at the back of the container.

4 Plant a verbena on either side of the geranium at the front of the window box. Water well and stand in a sunny position.

GARDENER'S TIP

White geraniums need regular dead-heading to look their best. Old flowerheads discolour and spoil the appearance of the plant.

Plant in late spring or early summer

Wedding Bells

A painted wooden window box filled with a white geranium, verbena, marguerites, *Bacopa* and silver *Senecio* is an ideal combination of plants for a summer wedding celebration in the garden.

GARDENER'S TIP

To prolong the life of wooden containers it is advisable to empty them of compost before winter and store them under cover until spring

Plant in late spring or early summer

MATERIALS
45 cm (18 in) slatted wooden window box
Sphagnum moss
Compost
Slow-release plant food granules

PLANTS
White geranium (*Pelargonium*)
2 white marguerites (*Argyranthemum*)
2 white trailing verbena
2 *Senecio cineraria* 'Silver Dust'
White *Bacopa*

1 It is a good idea to line slatted wooden containers with moss before planting to prevent the compost from leaking out.

2 Fill the moss-lined window box with compost, mixing in 2 teaspoons of slow-release plant food granules. Plant the geranium in the centre of the window box towards the back.

3 Plant the marguerites on either side of the geranium.

4 Plant the verbenas in the two back corners of the window box.

5 Plant the *Senecio* in the front two corners of the window box.

6 Plant the *Bacopa* centrally in the front of the box. Water thoroughly and stand in a sunny position.

White Flowers for Summer Evenings

The papery white flowers of the petunias are underplanted with white lobelia and surrounded by silver *Helichrysum* and the delicate daisy flowers of the *Erigeron*. This basket looks wonderful in the pale light of a summer's evening – hang it near a table as a decoration for alfresco dining.

MATERIALS
30 cm (12 in) hanging basket
Sphagnum moss
Compost
Slow-release plant food granules

PLANTS
4 white lobelia
3 white petunia
Helichrysum microphyllum
3 *Erigeron mucronatus*

lobelia

petunia

Erigeron

Helichrysum

1 Half line the basket with moss and position the lobelias around the side of the basket. Rest the rootballs on the moss and gently feed the foliage through the side of the basket.

2 Finish lining the basket with moss and fill with compost. Mix a teaspoon of slow-release plant food granules into the top of the compost. Plant the petunias in the top of the basket.

3 Plant the *Helichrysum* in the middle of the basket between the petunias.

4 Plant the *Erigeron* daisies between the petunias. Water well and hang in a sunny position.

GARDENER'S TIP

Light up summer parties with night-lights in votive glasses tucked into the hanging baskets for a magical effect. Make sure that stray foliage and flowers are pushed out of the way.

Plant in late spring or early summer

An Informal Wall Basket in Silver, White and Pink

The strong pink of the dahlietta flower is echoed in the leaf colouring of the pink-flowered *Polygonum* in this country-style basket. Silver-leaved thyme and white lobelia provide a gentle contrast.

MATERIALS
36 cm (14 in) wall basket
Sphagnum moss
Compost
Slow-release plant food granules

PLANTS
5 white lobelia
3 *Polygonum* 'Pink Bubbles'
2 *Thymus* 'Silver Queen'
1 pink dahlietta (miniature dahlia)

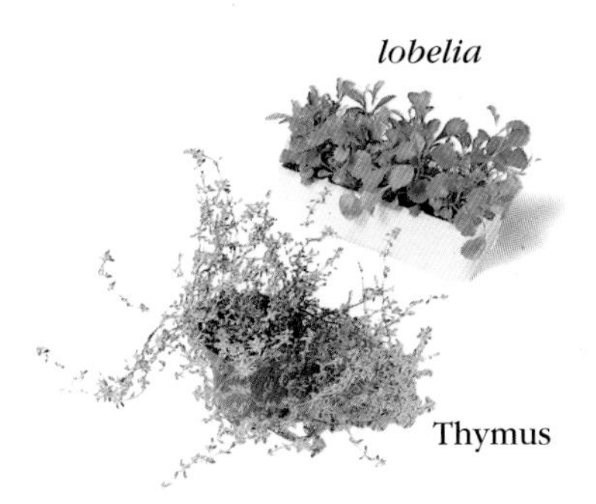

GARDENER'S TIP

To prevent the thyme getting leggy, trim off all the flowerheads after flowering - this will help maintain a dense, well-shaped plant.

Plant in spring

1 Line the back and the base of the basket with moss and position three lobelias around the side of the basket near the base.

2 Plant two of the *Polygonum* into the side of the basket above the lobelia. Rest the rootballs on the moss and gently feed the foliage through the side of the basket.

3 Fill the basket with compost. Mix a half-teaspoon of slow-release plant food granules into the top of the compost. Plant the thymes into the corners of the basket, angling them so that they tumble over the sides.

4 Plant the dahlietta in the middle of the basket and the remaining *Polygonum* in front of the dahlietta. Plant the remaining lobelias. Water well and hang in a sunny position.

Topiary Ivy with White Petunias

Use wire topiary frames (available at most garden centres) to train ivy or other climbing plants into interesting shapes. The ivy will take some months to really establish its outline; in the meantime, miniature white petunias complete the picture.

Gardener's Tip

Maintain the shape of the ivy with regular trimming and training – five minutes once a week will create a better shape than 15 minutes once a month.

Plant ivy at any time of year, petunias in spring

Materials
45 cm (18 in) oval terracotta window box
Crocks or other suitable drainage material
Compost
Slow-release plant food granules
Wire topiary frame
Pins made from garden wire
Plant rings

Plants
2 variegated ivies
4 miniature white petunias

1 Place a layer of drainage material in the base of the window box. Fill the window box with compost, mixing in 2 teaspoons of slow-release plant food granules.

2 Plant the two ivies, one in front of the other in the centre of the window box.

3 Position the topiary frame in the centre of the window box and use pins to hold it in place.

4 Wrap the stems of ivy around the stem of the frame and then around the frame itself.

5 Cut away any straggly stems and use plant rings to secure the ivy to the frame.

6 Plant the petunias around the topiary ivy. Water thoroughly and stand in light shade.

A Basket of Pinks

In this basket-weave stone planter sugar-pink petunias are planted with ivy-leaved geraniums and shaggy-flowered pink *Dianthus* with a deep-red eye. None of these plants requires much depth for its roots and provided the plants are fed and watered regularly they will be perfectly happy.

Gardener's Tip

Once the summer is over, the petunias and geraniums will need to be removed, but the *Dianthus* will overwinter quite happily. Cut off any flower stems and add a fresh layer of gravel.

Plant in late spring or early summer

MATERIALS
60 cm (24 in) window box
Washed gravel
Compost
Slow-release plant food granules

PLANTS
2 pink-flowered ivy-leaved geraniums (*Pelargoniums*)
3 sugar-pink petunias
6 pink *Dianthus*

geranium

petunia

Dianthus

1 Fill the base of the window box with a layer of gravel or similar drainage material.

2 Fill the window box with compost, mixing in 2 teaspoons of slow-release plant food granules.

3 Plant the two geraniums about 10 cm (4 in) from either end of the window box.

4 Plant the petunias, evenly spaced, along the back edge of the window box.

5 Plant four of the *Dianthus* along the front edge of the window box and the other two plants on either side of the central petunia.

6 Spread a layer of gravel around the plants; this is decorative and also helps to retain moisture. Water well and stand in a sunny position.

A Pretty Stencilled Planter

This small stencilled wooden window box is full of blue flowers. In a particularly pretty mix, petunias are intertwined with *Brachycome* daisies and trailing *Convolvulus*. A pair of brackets hold it in place under the window.

MATERIALS
40 cm (16 in) wooden window box (stencilling optional)
Clay granules or other suitable drainage material
Compost
Slow-release plant food granules

PLANTS
3 blue petunias
2 blue *Brachycome* daisies
Convolvulus sabatius

1 Line the base of the window box with clay granules or other suitable drainage materials.

2 Fill the window box with compost, mixing in a teaspoon of slow-release plant food granules. Plant the three petunias, evenly spaced, towards the back of the window box.

3 Plant the *Brachycome* daisies between the petunias.

4 Plant the *Convolvulus* centrally at the front of the box. Water thoroughly and position in full or partial sun.

GARDENER'S TIP

If you are stencilling a wooden container for outside use, do not forget to seal the wood after decorating it. In this instance, a matt wood varnish in a light oak tint has been used.

Plant in late spring or early summer

Regency Lily Urn

The shape of this urn is based on the shape of the lily flower, so it makes an appropriate container for this lovely mix of lilies, lavenders, pink marguerites and *Helichrysum*.

Materials and Tools
Suitably shaped urn
Gravel
Loam-based compost with
1/3 added grit
Slow-release plant food granules
Trowel

Plants
2 white lilies
2 dwarf lavender ('Hidcote' was used here)
2 pink marguerites
3 ***Helichrysum petiolatum***

lily

lavender

marguerite

Helichrysum petiolatum

1 Place a 5 cm (2 in) layer of gravel at the bottom of the urn and half-fill the container with compost. Place the lilies in the centre of the container.

2 Arrange the lavenders and marguerites around the lilies.

3 Plant the *Helichrysum* around the edge of the urn so that they can cascade over the rim as they grow. Fill between the plants with additional compost enriched with a tablespoon of slow-release plant food granules. Water well and place in a sunny position.

Gardener's Tip

Cut back the lavender heads when they have finished flowering, leave the lilies to die down naturally and dead-head the marguerites regularly to keep them flowering all summer. The ***Helichrysum*** and marguerites are not frost hardy, but the lavender and lilies should bloom again next year.

Plant in spring to flower in summer.

In the Pink

The common name for *Dianthus deltoides* is "the Pink". Its delightful deeply coloured flowers and silvery grey foliage work very well in a hanging basket combined with prostrate thymes, pink-flowered verbena and an *Osteospermum*.

GARDENER'S TIP

Pinch out the growing tips regularly to prevent plants such as the *Osteospermum* growing too vigorously upwards and unbalancing the look of the basket. It will be bushier and more in scale with the other plants as a result.

Plant in spring

MATERIALS
36 cm (14 in) hanging basket
Sphagnum moss
Compost
Slow-release plant food granules

PLANTS
6 pinks (*Dianthus deltoides*)
Osteospermum 'Pink Whirls'
Verbena 'Silver Anne'
3 thyme (*Thymus*) 'Pink Chintz' or similar prostrate variety

1 Line the bottom half of the basket with moss and fill with compost.

2 Plant three of the pinks into the side of the basket, resting the rootballs on the compost and feeding the leaves carefully through the wire.

3 Line the rest of the basket with moss and fill with compost. Mix a teaspoon of slow-release plant food granules into the top of the compost. Plant the *Osteospermum* in the centre of the basket.

4 Plant the verbena to one side of the *Osteospermum* on the edge of the basket.

5 Plant the thymes evenly spaced around the unplanted edge of the basket.

6 Plant the remaining three pinks between the thymes and the verbena. Water well and hang in a sunny position.

A Wall Basket in Shades of Pink

Trailing rose-pink petunias provide the main structure of this wall basket and are combined with two colourful verbenas and white alyssum. On their own, the pale petunia flowers could look somewhat insipid but they are enhanced by the deeper tones of the verbenas.

MATERIALS
36 cm (14 in) wall basket
Sphagnum moss
Compost
Slow-release plant food granules

PLANTS
4 white alyssum
2 cascading rose-pink petunias
2 *Verbena* 'Pink Parfait' and 'Carousel', or similar

GARDENER'S TIP

If, like these petunias, some of the plants are more developed than the others, pinch out the growing tips so that all the plants develop together and one variety will not smother the others.

Plant in late spring or early summer

1 Line the back of the basket and half-way up the front with moss.

2 Plant the alyssum into the side of the basket, resting the rootballs on the moss and feeding the foliage through the sides.

3 Fill the basket with compost and mix a half-teaspoon of slow-release plant food granules into the top of the compost. Plant the petunias in either corner of the basket.

4 Plant the verbenas, one in front of the other, in the middle of the basket. Water thoroughly and hang in a sunny position.

'Balcon' Geraniums

Traditionally planted to cascade from the balconies of houses and flats in many European countries, these lovely geraniums are now increasingly and deservedly popular. They are seen at their best when planted alone, as in the basket shown here, where the only variation is of colour.

MATERIALS
40 cm (16 in) hanging basket
Sphagnum moss
Compost
Slow-release plant food granules

PLANTS
5 'Balcon' geraniums (*Pelargonium* 'Princess of Balcon' and 'King of Balcon' were used here)

'Balcon' geraniums

GARDENER'S TIP

Take cuttings from non-flowering stems in the autumn to use in next year's basket. Geranium cuttings root easily and the young plants can be kept on a windowsill until spring.

Plant in late spring or early summer

1 Fully line the basket with moss.

2 Fill with compost. Mix a teaspoon of slow-release plant food granules into the top layer of the compost.

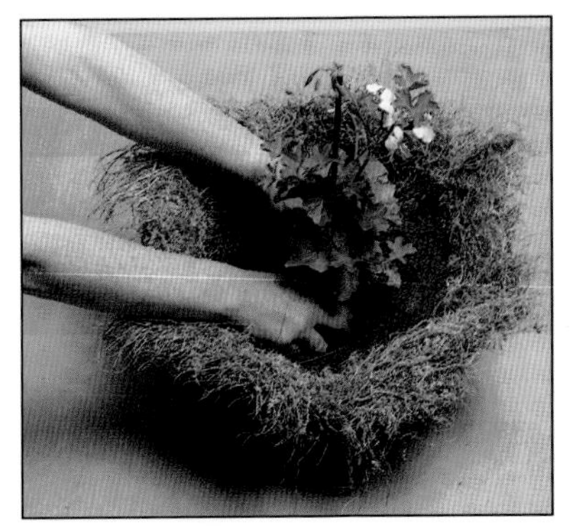

3 Plant one of the geraniums in the centre of the basket.

4 Plant the other four geraniums round the edge of the basket and remove any supporting canes to encourage the plants to tumble over the side. Water well and hang in a sunny spot.

A Small Basket of Geraniums

Ivy-leaved geraniums (*Pelargoniums*) are lovely plants for hanging baskets and one plant will fill a small basket like this by the middle of summer. The silver-leaved *Helichrysum* and lilac *Diascia* add the finishing touches to a pink-and-silver theme.

MATERIALS
25 cm (10 in) hanging basket
Sphagnum moss
Compost
Slow-release plant food granules

PLANTS
2 *Diascia* 'Lilac Belle'
Ivy-leaved geranium (*Pelargonium*) 'Super Rose'
2 *Helichrysum microphyllum*

1 Line the bottom half of the hanging basket with moss.

2 Plant the *Diascia* into the side of the basket by resting the rootballs on the moss and gently feeding the foliage between the wires. Add some compost.

3 Line the rest of the basket with moss, top up with compost and mix a teaspoon of slow-release plant food granules into the top layer. Plant the geranium in the centre of the basket.

4 Plant the *Helichrysum* on either side of the geranium. Water well and hang in a sunny position.

GARDENER'S TIP

If you like some height in your hanging basket, use small canes to support some of the geraniums stems; if you prefer a cascading effect, leave the geraniums unsupported.

Plant in late spring or early autumn

Hot Flowers in a Cool Container

Shocking-pink petunias and verbenas are the dominant plants in this window box which also features softer pink marguerites and silver *Helichrysum*. The dark green of the wooden window box is a calming influence which contrasts pleasingly with the vibrant flowers.

MATERIALS
76 cm (30 in) plastic window box
90 cm (3 ft) wooden window box (optional)
Compost
Slow-release plant food granules

PLANTS
Trailing pink marguerite (*Argyranthemum*) 'Flamingo'
2 bright pink verbenas, such as 'Sissinghurst'
3 shocking-pink petunias
4 *Helichrysum petiolare microphyllum*

Plant in late spring or early summer

1 Check the drainage holes are open in the base and, if not, drill or punch them open. Fill the window box with compost, mixing in 3 teaspoons of slow-release plant food granules. Plant the marguerite centre front.

2 Plant the verbenas in the back corners of the window box.

3 Plant one of the petunias behind the marguerite and the other two on either side of it.

4 Plant one *Helichrysum* on each side of the central petunia and the other two *Helichrysum* in the front corners of the window box. Water well and lift into place in the wooden window box, if using. Stand in a sunny position.

A Peachy Pink Wall Basket

The vivid petunia and geranium (*Pelargonium*) flowers contrast dramatically with the greeny-yellow of the lamium. This basket is best against a dark background.

MATERIALS
30 cm (12 in) wide wall basket
Sphagnum moss
Compost
Slow-release plant food granules

PLANTS
3 *Lamium* 'Golden Nuggets'
Peach/pink geranium (zonal *Pelargonium*) 'Palais', or similar
3 petunias

1 Line the back and lower half of the front of the basket with moss.

2 Plant the *Lamium* into the side of the basket by resting the rootball on the moss and feeding the foliage through the side of the basket. Line the rest of the basket with moss.

3 Fill the basket with compost, mixing a half-teaspoon of slow-release plant food granules into the top layer. Plant the geranium (*Pelargonium*) in the centre of the basket against the back edge.

4 Plant one petunia in each corner and the third in front of the geranium. Water well and hang on a sunny wall.

GARDENER'S TIP

For a gentler colour scheme, the *Lamium* can be replaced with the silver-grey foliage of *Helichrysum microphyllum*.

Plant in late spring or early summer

An Instant Garden

There is not always time to wait for a window box to grow and this is one solution. Fill a container with potted plants and, as the season progresses, you can ring the changes by removing those that are past their best and introducing new plants.

MATERIALS
64 cm (25 in) galvanized tin window box
Clay granules
5 1-litre (4 in) plastic pots
Compost

PLANTS
Lavandula pinnata
2 blue petunias
Convolvulus sabatius
Blue *Bacopa*
Helichrysum petiolare
Viola 'Jupiter', or similar

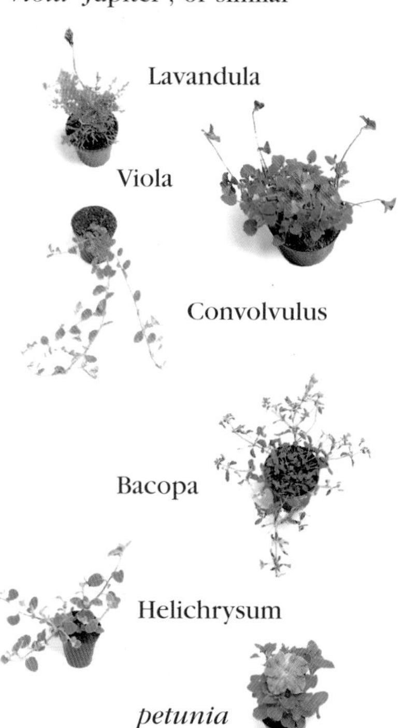

1 Fill the base of the container with clay granules or similar drainage material.

2 Pot up the lavender into one of the pots.

3 Pot up one of the petunias with the *Convolvulus*.

4 Pot up the other petunia with the *Bacopa*.

5 Pot up the *Helichrysum*.

6 Pot up the viola, and arrange the pots in the window box.

GARDENER'S TIP

When using a container without drainage holes, take care not to overwater or the roots will become waterlogged. Check after heavy rain, too, and empty away any excess water.

Plant in late spring or early summer

A Cascade of Blue and Silver

Blue petunias and violas are surrounded by a cascading curtain of variegated ground ivy and silver-leaved *Senecio* in this softly coloured hanging basket.

MATERIALS
30 cm (12 in) hanging basket
Sphagnum moss
Compost
Slow-release plant food granules

PLANTS
3 deep blue violas
3 soft blue petunias
Variegated ground ivy (*Glechoma hederacea* 'Variegata')
3 *Senecio cineraria* 'Silver Dust'

1 Line the lower half of the basket with moss.

3 Line the rest of the basket with moss and fill with compost, mixing a teaspoon of slow-release plant food granules into the top layer.

2 Plant the violas in the side of the basket by resting the rootballs on the moss and carefully guiding the foliage between the wires.

4 Plant the three petunias, evenly spaced, in the top of the basket.

5 Plant the ground ivy on one side so that it trails over the edge of the basket.

GARDENER'S TIP

If the ground ivy becomes too rampant and threatens to throttle the other plants, prune it by removing some of the stems completely and reducing the length of the others.

Plant in late spring or early summer.

6 Plant the *Senecio* plants between the petunias. Water well and hang in a sunny position.

Violas and Verbena

Deep blue violas are surrounded by trailing purple verbena to make a simple but attractive basket. Trailing verbena is a particularly good hanging-basket plant with its feathery foliage and pretty flowers.

MATERIALS
30 cm (12 in) hanging basket
Sphagnum moss
Compost
Slow-release plant food granules

PLANTS
9 blue violas
3 purple trailing verbenas

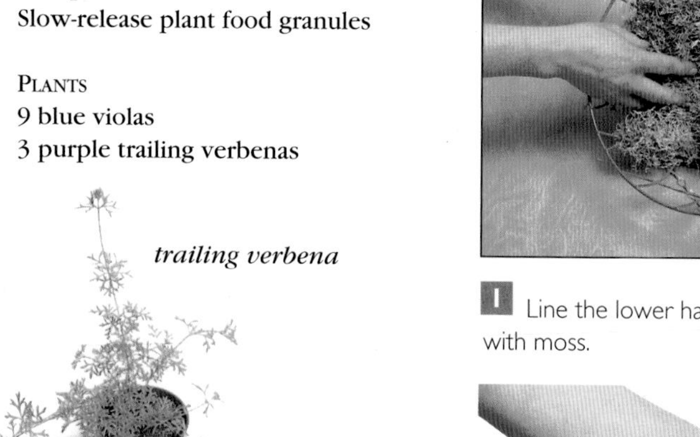

trailing verbena

violas

1 Line the lower half of the basket with moss.

2 Plant five of the violas into the side of the basket by resting the rootballs on the moss and guiding the foliage through the side of the basket.

3 Line the rest of the basket with moss and fill with compost, mixing a teaspoon of slow-release plant food granules into the top layer. Plant the verbenas around the edge of the basket

4 Plant the remaining violas in the centre of the basket. Water well and hang in partial sun.

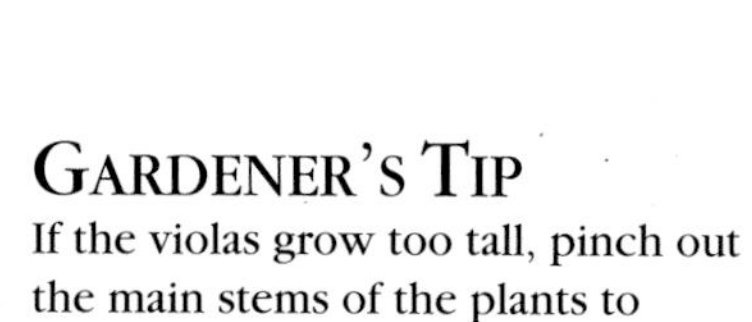

GARDENER'S TIP

If the violas grow too tall, pinch out the main stems of the plants to encourage the spreading side shoots.

Plant in spring

Shades of Blue

Some unlikely plants, such as this powder-blue scabious, can do very well in a hanging basket, especially when combined, as it is here, with *Isotoma* and *Ageratum* in the same colour and the trailing silver foliage of *Helichrysum*.

MATERIALS
40 cm (16 in) hanging basket
Sphagnum moss
Compost
Slow-release plant food granules

PLANTS
6 blue *Ageratum*
Blue scabious
3 *Helichrysum petiolare*
3 blue *Isotoma axillaris*

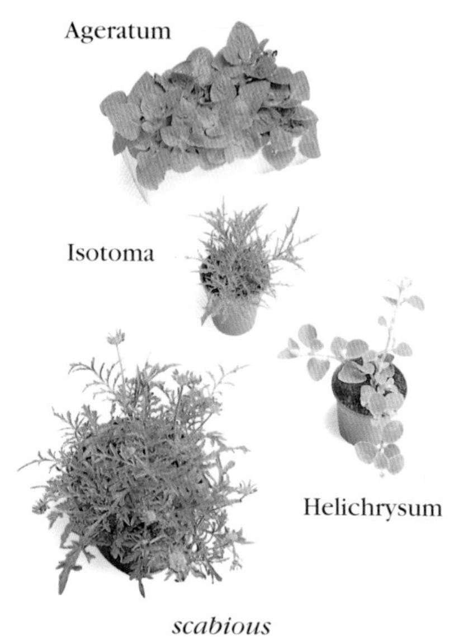

1 Line half of the basket with moss.

2 Plant three of the *Ageratum* into the side of the basket by resting the rootballs on the moss and carefully guiding the foliage through the side.

3 Add a further layer of moss and plant the other three *Ageratum* into the side of the basket at a higher level.

4 Fill the basket with compost, mixing a teaspoon of slow-release plant food granules into the top layer. Plant the scabious in the centre of the basket.

5 Plant the *Helichrysum* evenly spaced around the edge of the basket.

6 Plant the *Isotoma* between the *Helichrysum*. Water well and hang in a sunny position.

GARDENER'S TIP

At the end of the season the scabious can be removed from the basket and planted in the border to flower for many years to come.

Plant in late spring

Showers of Flowers

Deep, velvety purple pansies and purple sage are surrounded by pink *Nemesia* and tumbling purple-pink verbena in a pretty basket hung here in the corner of a thatched summerhouse.

MATERIALS
40 cm (16 in) hanging basket
Sphagnum moss
Compost
Slow-release plant food granules

PLANTS
3 purple verbenas
Purple sage
3 deep purple pansies (*Viola*)
6 *Nemesia* 'Confetti'

1 Line the lower half of the basket with moss.

2 Plant the verbenas in the side of the basket. Line the basket with moss and fill with compost, mixing a teaspoon of plant food granules into the compost.

3 Plant the sage in the middle of the basket. Then plant the three purple pansies around the sage.

4 Add more compost around the pansies and press in firmly.

5 Plant three *Nemesia* at the back of the pansies.

6 Plant the remaining *Nemesia* between the pansies. Water thoroughly and hang in light shade or partial sun.

GARDENER'S TIP

In summer, pansies tend to flag in hot sun, especially when planted in hanging baskets. They will do best where they are in the shade during the hottest part of the day.

Plant in spring

Flame-red Flowers in Terracotta

The intense red flowers of the geraniums, verbenas and nasturtiums are emphasized by a few yellow nasturtiums and the variegated ivy, but cooled slightly by the soothing blue-green of the nasturtium's umbrella-shaped leaves.

GARDENER'S TIP

Nasturtiums are prone to attack by blackfly. Treat at the first sign of infestation with a suitable insecticide and the plants will remain healthy.

Plant in late spring or early summer

MATERIALS

50 cm (20 in) terracotta window box
Crocks or other suitable drainage material
Compost
Slow-release plant food granules

PLANTS

2 red geraniums (zonal *Pelargoniums*)
2 nasturtiums - 1 red, 1 yellow
Red verbena
2 variegated ivies

1 Place a layer of drainage material in the base of the window box.

2 Fill the container with compost, mixing in 3 teaspoons of slow-release plant food granules.

3 Plant the geraniums either side of the centre of the window box.

4 Plant a nasturtium at either end, in the back corners.

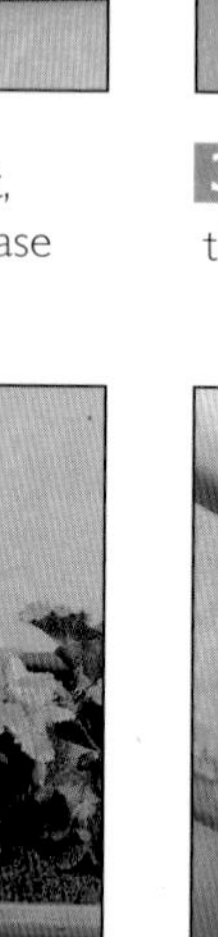

5 Plant the verbena in the centre of the window box.

6 Plant the ivies in front of the nasturtiums in the corners. Water well, leave to drain, and place in a sunny position.

A Sunny Wall Basket

The vibrant yellows, oranges and reds of the flowers in this basket glow richly amongst the variegated leaves of the nasturtiums. As the season progresses the underplanted *Lysimachia* will bear deep yellow flowers and add another layer of colour to this basket.

MATERIALS
30 cm (12 in) wall basket
Sphagnum moss
Compost
Slow-release plant food granules

PLANTS
2 *Lysimachia congestiflora*
3 nasturtium 'Alaska'
3 mixed colour African marigolds (*Tagetes*)

nasturtium

Lysimachia

African marigolds

1 Line the back of the basket and half-way up the front with moss.

2 Plant the *Lysimachia* into the side of the basket by resting the rootballs on the moss and carefully feeding the foliage between the wires.

3 Fill the basket with compost, mixing a half-teaspoon of slow-release plant food granules into the top layer. Plant the nasturtiums along the back of the basket.

4 Plant the African marigolds in front of the nasturtiums. Water well and hang in a sunny spot.

GARDENER'S TIP

If you have a large area of wall to cover, group two or three wall baskets together. This looks very effective, especially when they are planted with the same plants.

Plant in spring

Mediterranean Mood

The *Lantana* is a large shrub which thrives in a Mediterranean or sub-tropical climate, but it is increasingly popular in cooler climates as a half-hardy perennial in borders and containers. This multi-coloured variety has been planted with yellow *Bidens* and orange dahliettas.

GARDENER'S TIP

To encourage a bushy plant, pinch out the growing tips of the *Lantana* regularly. Like many popular plants, the *Lantana* is poisonous, so treat it with respect and do not try eating it!

Plant in late spring or early summer

MATERIALS
36 cm (14 in) hanging basket
Sphagnum moss
Compost
Slow-release plant food granules

PLANTS
Orange/pink *Lantana*
3 orange dahliettas (miniature dahlia)
3 *Bidens aurea*

1 Line the basket with moss. Fill the basket with compost, mixing a teaspoon of slow-release plant food granules into the top layer. Plant the *Lantana* in the centre of the basket.

2 Plant the dahliettas, evenly spaced, around the *Lantana*.

3 Plant the *Bidens* between the dahliettas. Water thoroughly and hang in a sunny position.

Sunny Daisies with Violas and Bacopa

Osteospermum daisies are sun-worshippers, keeping their petals furled in cloudy weather. In this window box they are combined with yellow violas and tumbling white *Bacopa*.

MATERIALS
45 cm (18 in) fibre window box
Polystyrene or other suitable drainage material
Compost
Slow-release plant food granules

PLANTS
Osteospermum 'Buttermilk'
3 yellow violas
2 white *Bacopa*

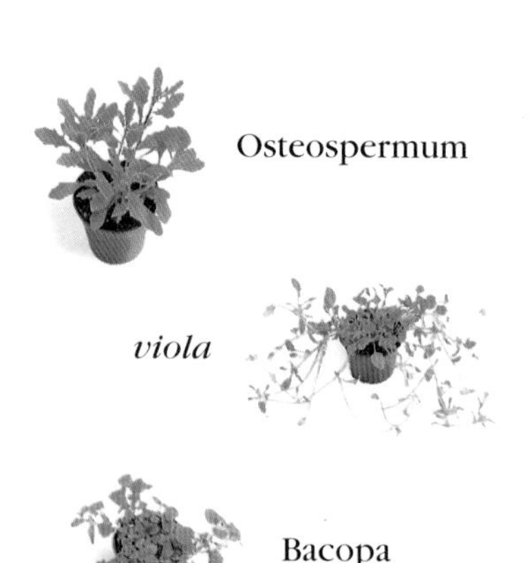

Osteospermum

viola

Bacopa

1 Line the base with polystyrene or other suitable drainage material.

2 Fill the window box with compost, mixing in 3 teaspoons of slow-release plant food granules. Plant the *Osteospermum* in the centre of the window box.

3 Plant two of the violas at either end of the window box and the third in front of the *Osteospermum*.

4 Plant the two *Bacopa* on either side of the *Osteospermum*. Stand in a sunny spot and water thoroughly.

GARDENER'S TIP

Pinch out the growing tips of the *Osteospermum* regularly to encourage a bushy rather than a leggy plant.

Plant in spring

Alaska Nasturtiums with Snapdragons and Daisies

The leaves of the Alaska nasturtium look as if they have been splattered with cream paint. In this window box they are planted with yellow-flowered snapdragons, *Gazania* and *Brachycome* daisies.

MATERIALS
76 cm (30 in) plastic window box
Compost
Slow-release plant food granules

PLANTS
2 yellow *Gazania*
3 Alaska nasturtiums
3 *Brachycome* 'Lemon Mist'
2 yellow snapdragons (*Antirrhinum*)

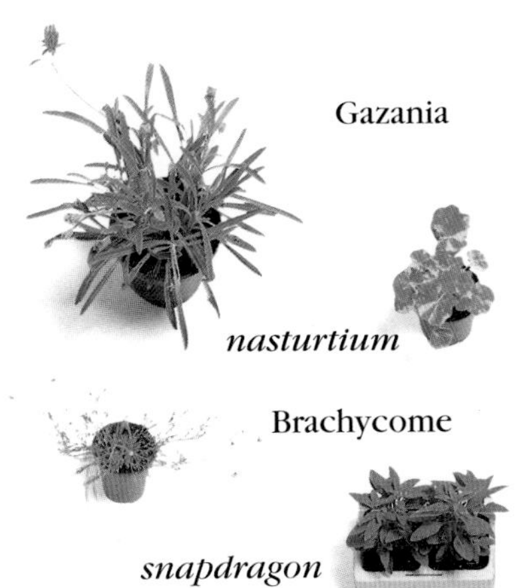

GARDENER'S TIP

Nasturtiums are amongst the easiest plants to grow from seed. Start them off about 4 to 6 weeks before you plant your window box, potting them on to keep them growing vigorously.

Plant in the spring

1 Drill drainage holes in the window box, if necessary. Fill the window box with compost, mixing in 2 teaspoons of slow-release plant food granules. Plant the two *Gazania*, evenly spaced half-way from the centre.

2 Plant the nasturtiums at either end and in the centre of the window box.

3 Plant the three *Brachycome* daisies, evenly spaced along the front of the window box.

4 Plant the two snapdragons either side of the central nasturtium. Water thoroughly, leave to drain, and stand in a sunny position.

A Touch of Gold

Yellow *Lantana* and the yellow-flowered variegated-leaf nasturtium provide colour from early summer onwards and later in the season the black-eyed Susan will be covered in eye-catching flowers.

MATERIALS
30 cm (12 in) hanging basket
Sphagnum moss
Compost
Slow-release plant food granules

PLANTS
3 nasturtiums 'Alaska'
Yellow *Lantana*
3 black-eyed Susans (*Thunbergia alata*)

1 Line the lower half of the basket with moss.

2 Plant the nasturtiums into the side of the basket by resting the rootballs on the moss and carefully guiding the leaves through the side of the basket.

3 Complete lining the basket with moss. Fill the basket with compost, mixing a teaspoon of slow-release plant food granules into the top layer.

4 Plant the *Lantana* in the centre of the basket.

5 Plant the black-eyed Susans around the *Lantana*. Water well and hang in a sunny position.

GARDENER'S TIP

Save some of the nasturtium seeds for next year's baskets and pots – they are among the easiest of plants to grow.

Plant in late spring or early summer

A Floral Chandelier

The chandelier shape is a result of combining the spreading *Bidens* with upright *Lantana* and marigolds. The variegated-leaf *Lantana* proved very slow to establish so a more vigorous green-leaved variety was added later. As the season progresses, the strongly marked leaves of the variegated plants will become more dominant.

MATERIALS
36 cm (14 in) hanging basket
Sphagnum moss
Compost
Slow-release plant food granules

PLANTS
3 yellow *Lantana*, 2 variegated, 1 green-leaved
2 *Bidens ferulifolia*
5 African marigolds (*Tagetes*)

1 Line the basket with moss.

2 Fill the basket with compost, mixing a teaspoon of slow-release plant food granules into the top layer. Plant the *Lantana*.

3 Plant the *Bidens* opposite one another at the edge of the basket.

4 Plant the African marigolds around the *Lantana* plants. Water thoroughly and hang in a sunny position.

GARDENER'S TIP

To complete the chandelier, make candle holders by twisting thick garden wire around the base of yellow candles and add them to the hanging basket.

Plant in late spring or early summer

A Wall Basket of Tumbling Violas

Violas can be surprisingly vigorous plants and, given the space, will happily tumble over the edge of a wall basket. Combined with parsley and the daisy-like flowers of *Asteriscus,* the effect is delicate but luxuriant.

MATERIALS
30 cm (12 in) wall basket
Compost
Sphagnum moss
Slow-release plant food granules

PLANTS
5 parsley plants
5 yellow violas
Asteriscus 'Gold Coin'

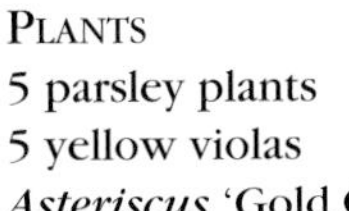

parsley

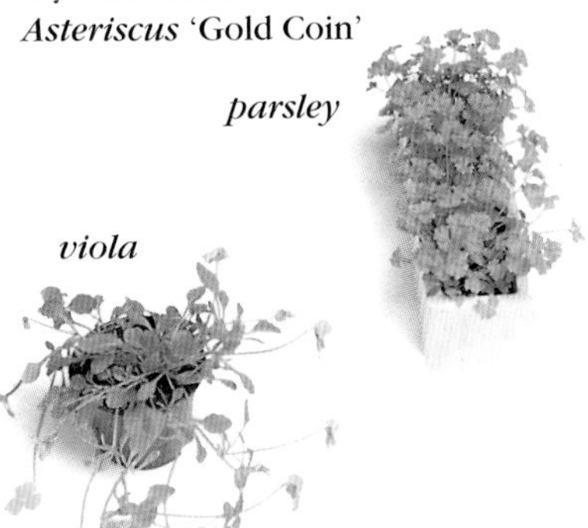

viola

Asteriscus

1 Line the back and lower half of the front of the basket with moss.

2 Plant three of the parsley plants into the sides of the basket by resting the rootballs on the moss and feeding the foliage through the wires.

3 Add another layer of moss and plant two of the viola plants in the wall of the basket using the same method.

4 Complete lining the basket with moss and fill with compost, mixing a half-teaspoon of slow-release plant food granules into the top layer. Plant the *Asteriscus* in the centre of the basket and surround with the remaining parsley and viola plants.

GARDENER'S TIP

To keep the violas flowering all summer they need regular dead-heading – the easiest way to do this is to give the plants a trim with a pair of scissors rather than trying to remove heads individually.

Plant in spring

A Miniature Cottage Garden

Part of this basket's charm is its simple planting scheme. Pot marigolds and parsley are planted with bright blue daisies to create a basket which would look at home on the wall of a cottage or outside the kitchen door.

MATERIALS
36 cm (14 in) hanging basket
Sphagnum moss
Compost
Slow-release plant food granules

PLANTS
5 parsley plants
3 pot marigolds (*Calendula*) 'Gitana', or similar
3 *Felicia*

Felicia

parsley

pot marigolds

1 Line the lower half of the basket with moss.

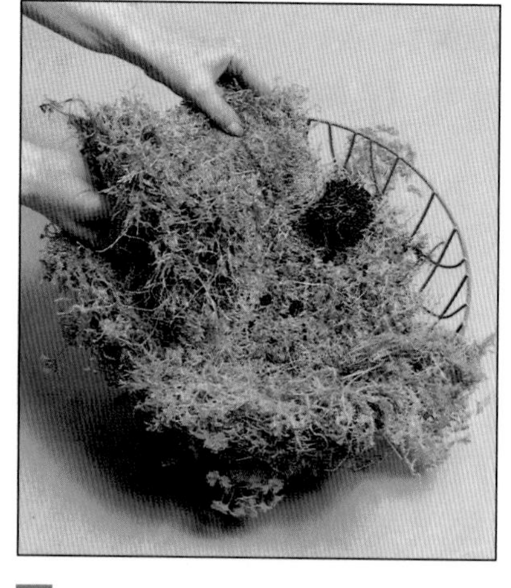

2 Plant the parsley into the sides of the basket by resting the rootballs on the moss and gently feeding the foliage through the sides.

3 Line the rest of the basket with moss, carefully tucking it around the roots of the parsley.

4 Fill the basket with compost, mixing a teaspoon of slow-release plant food granules into the top layer.

5 Plant the marigolds, evenly spaced in the top of the basket.

GARDENER'S TIP

Regular dead-heading will keep the basket looking good, but allow at least one of the marigold flowers to form a seed head and you will be able to grow your own plants next year.

Plant in spring

6 Plant the *Felicia* between the marigolds. Water well and hang in full or partial sun.

The Apothecary Box

Many plants have healing qualities and, while they should always be used with caution, some of the more commonly used herbs have been successful country remedies for centuries.

GARDENER'S TIP

Herbs should not be used to treat an existing medical condition without first checking with your medical practitioner.

Plant in the spring

MATERIALS
Wooden trug
Compost
2 tsp pelleted chicken manure or similar organic plant food

PLANTS
Lavender – for relaxing
Rosemary – for healthy hair and scalp
Chamomile – for restful sleep
Fennel – for digestion
Feverfew (*Matricaria*) – for migraine
3 pot marigolds (*Calendula*) – for healing

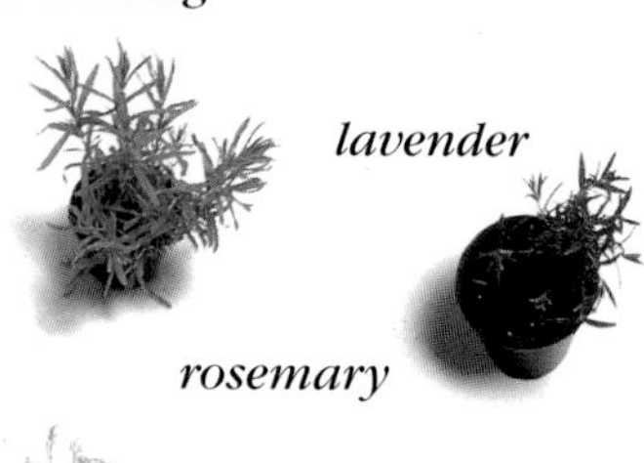

lavender
rosemary

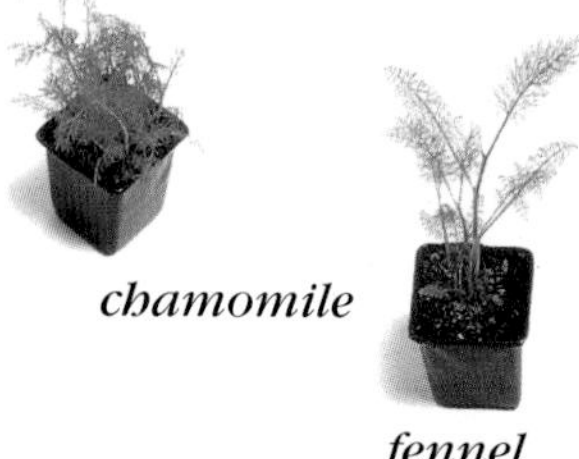

chamomile
fennel

feverfew

pot marigold

1 Place a layer of drainage material in the trug and fill with compost, mixing in 2 teaspoons of fertilizer.

2 Plant the lavender in the centre of the trug and the rosemary in the front right-hand corner.

3 Plant the chamomile in the back left-hand corner.

4 Plant the fennel in the back right-hand corner.

5 Plant the feverfew in the front left-hand corner.

6 Plant the marigolds in the spaces between the other herbs. Water well and stand in full or partial sun.

Herbs in the Shade

Although the Mediterranean herbs need lots of sunshine, there are many others which prefer a cooler situation to look and taste their best. This window box, ideal for just outside the kitchen door or window, has an interesting variety of mints, sorrel, chives, lemon balm and parsley.

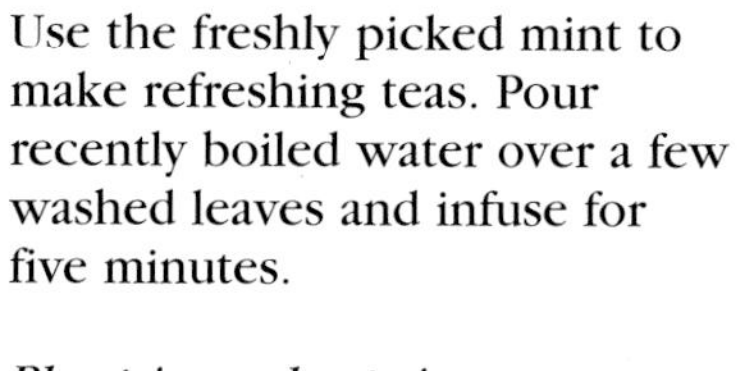

GARDENER'S TIP

Use the freshly picked mint to make refreshing teas. Pour recently boiled water over a few washed leaves and infuse for five minutes.

Plant in early spring

MATERIALS
- 50 cm (20 in) terracotta window box
- Polystyrene or other suitable drainage material
- Compost
- Slow-release plant food granules, pelleted chicken manure, or similar organic plant food

PLANTS
- Lemon balm (*Melissa officinalis*)
- 3 mints (black peppermint, silver mint and curly spearmint were used here)
- Sorrel
- Chives
- Parsley

1 Fill the base of the container with polystyrene or similar suitable drainage material. Fill the window box with compost, mixing in 3 teaspoons of slow-release plant food granules or organic alternative.

2 Plant the lemon balm in the back right-hand corner.

3 Plant two of the mints along the back of the window box and the third in the front right-hand corner of the container.

4 Plant the sorrel in the middle of the window box at the front.

5 Plant the chives at the front in the left-hand corner.

6 Finally, plant the parsley between the sorrel and the mint. Position in light shade. Water thoroughly.

Wine Case Herb Garden

Add a coat of varnish to an old wine case and make an attractive and durable container for a miniature herb garden. The container can be placed near the kitchen door or on the balcony.

MATERIALS AND TOOLS
Wooden wine case
Pliers
Sandpaper
Paintbrush
Light oak semi-matt varnish
Crocks or similar drainage material
Standard compost with ⅓ added coarse grit
Slow-release plant food granules
Bark mulch
Trowel

PLANTS
Selection of 7 herbs, such as sage, chives, parsley, mint, tarragon, lemon thyme and creeping thyme

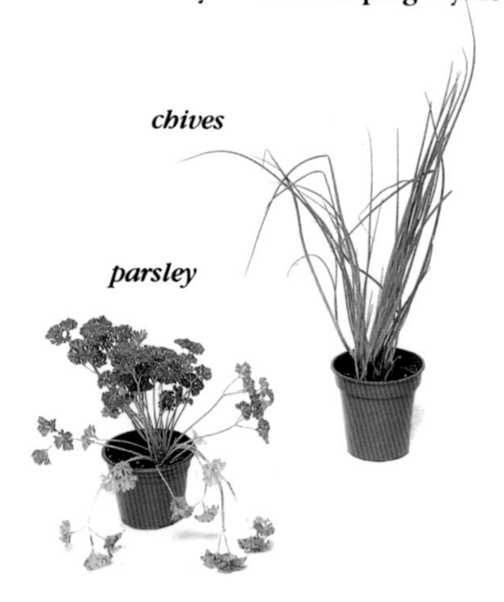

GARDENER'S TIP

Some herbs like cool, partial shade while others like hot, dry free-draining soil. A mixed herb garden will only thrive for one growing season.

Plant in spring.

1 Remove any wire staples from around the edges of the box and sand down any rough edges. Apply two coats of varnish on the inside and outside of the box, allowing the varnish to dry thoroughly between coats. Cover the base of the box with a layer of crocks or similar drainage material.

2 Before planting, plan how you are going to arrange the plants in the container to achieve a pleasing balance of colour, height and shape.

3 Fill the box with gritty compost and, working from one end of the box to the other, begin planting. Loosen the rootballs before planting, as this will help the plants to root into the surrounding compost.

4 Add the remaining plants. Scatter 2 tablespoons of slow-release plant food granules on the surface of the compost. Firm in the plants and mulch with a layer of bark, to retain moisture and prevent soil splashing the foliage. Water well.

Fruit and Flowers

Bright red petunias become even more vibrant when interplanted with variegated *Helichrysum* and underplanted with alpine strawberries. With their delicate trailing tendrils, the strawberry plants soften the lower edge of the basket.

MATERIALS
30 cm (12 in) hanging basket
Sphagnum moss
Compost
Slow-release plant food granules

PLANTS
3 alpine strawberry plants
3 bright red petunias
3 *Helichrysum petiolare* 'Variegatum'

Helichrysum

alpine strawberry

petunia

1 Line the lower half of the basket with moss.

2 Plant the alpine strawberries into the side of the basket by resting the rootballs on the moss and carefully guiding the leaves through the side of the basket.

3 Line the rest of the basket with moss and fill with compost. Mix a teaspoon of slow-release plant food granules into the top layer of compost. Plant the three petunias, evenly spaced in the top of the basket.

4 Interplant the petunias with the *Helichrysum*. Water thoroughly and hang in full or partial sun.

GARDENER'S TIP

The tendrils, or runners, sent out by the alpine strawberries are searching for somewhere to root. If you can fix a couple of pots to the wall, or the tendrils reach the ground, simply pin the plantlet into the compost or soil while it is still attached to the parent plant. As soon as it has rooted it can be cut free.

Plant in late spring or early summer

Good Enough to Eat

All the plants in this basket bear an edible crop; the tomato fruit, nasturtium flowers and parsley leaves. You could even impress your family or guests with a "hanging basket salad", using all three as ingredients.

MATERIALS
36 cm (14 in) hanging basket
Sphagnum moss
Compost
Slow-release plant food granules

PLANTS
6 parsley plants
5 trailing nasturtiums
3 tomatoes 'Tumbler', or similar

1 Line the lower half of the basket with moss.

2 Plant three parsley plants into the side of the basket by resting the rootballs on the moss and feeding the leaves through the side of the basket.

3 Put moss nearly to the lip of the basket and fill with compost. Mix a teaspoon of slow-release plant food granules into the top of the compost. Plant three nasturtium plants into the side of the basket, just below the lip.

4 Complete lining the basket with moss, being careful to tuck plenty of moss around the nasturtiums.

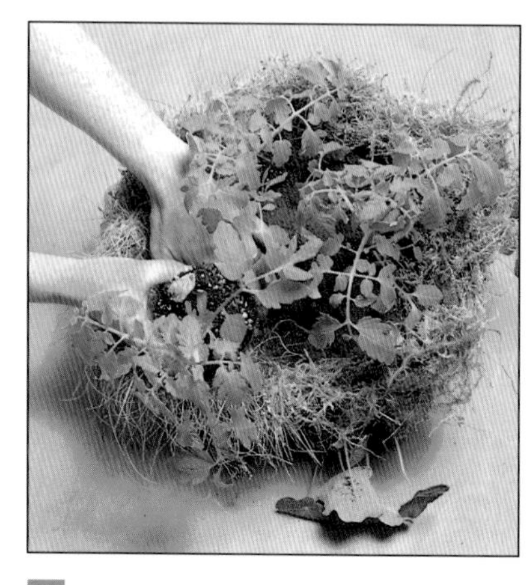

5 Plant the tomato plants in the top of the basket.

6 Plant the remaining three parsley plants amongst the tomatoes in the top of the basket. Water well and hang in a sunny position. Liquid feed regularly.

GARDENER'S TIP

If you would prefer to grow your plants organically, plant this basket in an organic compost and use natural plant foods such as pelleted chicken manure and a liquid seaweed feed.

Plant in late spring or early summer

Summer Fruits

The red of ripening strawberries is matched by vibrant red geranium (*Pelargonium*) flowers in this unusual arrangement. Alpine strawberry plants make a good contrast, with their delicate fruit and tendrils.

MATERIALS
36 cm (14 in) hanging basket
Sphagnum moss
Compost
Slow-release plant food granules

PLANTS:
3 strawberry plants 'Maxim', or similar
1 deep-red geranium (*Pelargonium*) 'Miss Flora'
3 alpine strawberry plants

alpine strawberry

strawberry

geranium

1 Line the basket with moss.

2 Fill the basket with compost. Mix a teaspoon of slow-release plant food granules into the top layer of compost. Plant the strawberry plants around the side of the basket.

3 Plant the geranium (*Pelargonium*) in the centre of the basket.

4 Plant the alpine strawberries in between the larger strawberry plants. Water well and hang in partial or full sun.

GARDENER'S TIP

After the large strawberries have fruited, cut back all the foliage on those plants to encourage the formation of next year's flowers. Keep well watered. The geranium and alpine strawberries will stop the basket looking too bare while the new foliage is growing.

Plant in spring

Potted Fruit Garden

Dwarf apple trees grow well in containers provided they are fed and watered regularly. With the addition of strawberry plants around the base of the apple you will be able to enjoy two crops of fruit from your potted garden. The strawberry plants are kept in their pots and bedded down in bark mulch as fruit trees don't like to compete for soil space.

MATERIALS AND TOOLS
Terracotta pot, 60 cm (24 in) high
Crocks or similar drainage material
Composted stable manure
Knife
Bonemeal
Loam-based compost
Bark mulch
Trowel

PLANTS
Apple tree on dwarf rooting stock
5 strawberry plants

strawberry plant

1 Cover the drainage holes in the base of the pot with crocks or similar drainage material and a 10 cm (4 in) layer of composted manure.

2 Use a knife to cut the polythene pot away from the tree rather than pulling the pot off, which can damage the rootball.

3 Place the tree in the pot and scatter a handful of bonemeal around the rootball. Then add the loam-based compost, filling the pot to approximately 10 cm (4 in) below the rim.

4 Place the strawberry plants in their pots around the trunk of the tree and fill the spaces in between with bark mulch. Place in a sunny position, and water and liquid feed regularly.

GARDENER'S TIP

Standard gooseberries and redcurrants also grow well in containers and look highly decorative when they are fruiting.

Plant in autumn or spring to fruit in summer.

The Good Life

This window box will not exactly make you self-sufficient, but it is surprising how many different vegetables can be grown in a small space. It is perfect for anyone who likes the taste of home-grown vegetables but does not have a garden to grow them in.

GARDENER'S TIP

It is now possible to buy "plugs" of small vegetable plants at garden centres. There is no need to separate the plants provided there is sufficient room between the clumps.

Plant in spring

MATERIALS
- 76 cm (30 in) plastic window box
- Compost
- Slow-release plant food granules, pelleted chicken manure or similar organic plant food

PLANTS
- Garlic
- 3 Chinese leaves
- 4 plugs of beetroot (see Gardener's Tip)
- Pepper plant
- 3 dwarf French beans
- 3 plugs of shallots

garlic

Chinese leaves

beetroot

pepper

dwarf French beans

shallots

1 Check the drainage holes are open in the base and, if not, drill or punch them open. Fill the window box with compost, mixing in 2 teaspoons of slow-release plant food granules or an organic alternative. Starting at the right-hand end of the window box, first plant the garlic.

2 Next plant the Chinese leaves.

3 Follow this with the plugs of beetroot.

4 Plant the pepper plant next, just to the left of the centre.

5 Now plant the three dwarf French bean plants.

6 Finally, plant the shallot plugs in the left-hand corner. Water well and stand in full or partial sun.

Vital Ingredients

A lovely present for an enthusiastic cook, this window box contains chervil, coriander, fennel, garlic, purple sage, French tarragon, savory, origanum and basil.

MATERIALS
45 cm (18 in) wooden window box
Crocks or other suitable drainage material
Compost
Slow-release plant food granules, pelleted chicken manure, or similar organic plant food

PLANTS
French tarragon
Chervil
Garlic
Coriander
Purple sage
Basil
Fennel
Savory (*Satureja hortensis*)
Origanum

1 Line the base of the container with crocks or other suitable drainage material. Fill the window box with compost, mixing in a teaspoon of slow-release plant food granules or organic alternative.

2 Before planting the herbs, arrange them in the window box in their pots.

Plant in spring

3 Plant the back row of herbs first.

4 Plant more herbs at the front of the window box. Water well and stand in a sunny position.

A Taste for Flowers

It is often said that something "looks good enough to eat" and in this instance it is true. All the flowers in this window box may be used for flavour and garnishes provided, of course, that they are washed.

MATERIALS
36 cm (14 in) terracotta window box
Crocks or other suitable drainage material
Compost
Slow-release plant food granules, pelleted chicken manure or similar organic plant food

PLANTS
Chives
2 nasturtiums
2 pansies with well-marked "faces"
2 pot marigolds (*Calendula*)

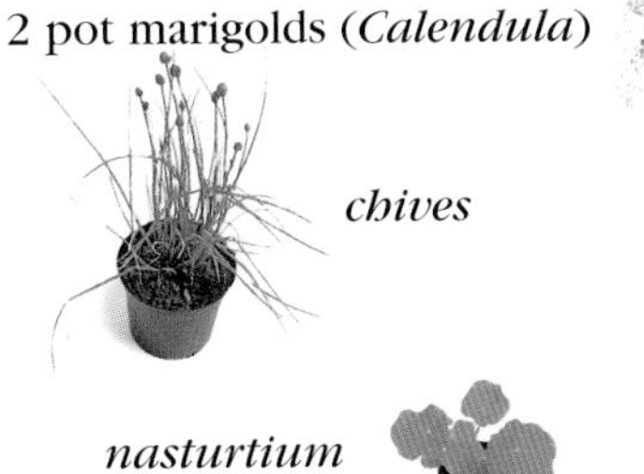

GARDENER'S TIP

To keep all the plants producing flowers, dead-head regularly. Once a plant has set seed it considers its work done and will produce fewer and fewer flowers.

Plant in early spring

1 Cover the base of the window box with a layer of crocks or similar drainage material. Fill with compost, mixing in 2 teaspoons of slow-release plant food granules or organic alternative. Plant the chives in the right-hand corner.

2 Plant one of the nasturtiums in the left-hand corner and the other centre front.

3 Plant one pansy at the back next to the chives and the other at the front to the left of the central nasturtium.

4 Plant one of the marigolds at the back between the pansy and nasturtium and the other one just behind the central nasturtium.

Sweet Scents for a Conservatory

A simple, very informal planting that would thrive in a conservatory. A pretty combination of scented-leaf geranium, deep-blue miniature petunias, purple trailing verbenas and exuberant tumbling ground ivy.

MATERIALS
30 cm (12 in) plastic window box
Compost
Slow-release plant food granules

PLANTS
Scented-leaf geranium (*Pelargonium*) 'Lady Plymouth'
2 variegated ground ivy (*Glechoma hederacea* 'Variegata')
2 deep-purple trailing verbena
2 deep-blue 'Junior' petunias

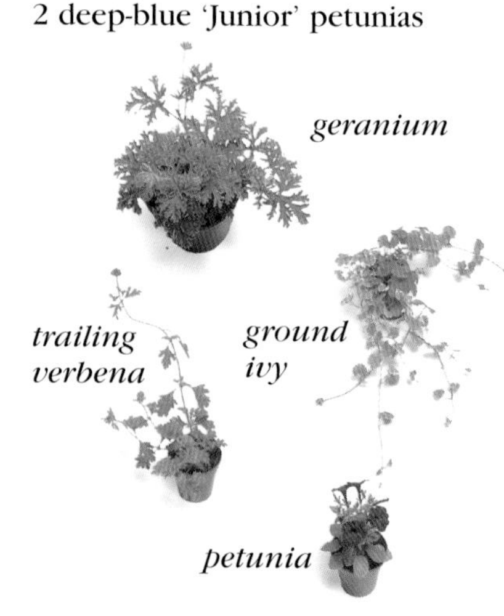

1 Check drainage holes are open; if not, drill or punch them out. Fill the window box with compost, mixing in a teaspoon of slow-release plant food granules. Plant the geranium in the centre.

2 Plant the ground ivies at either end of the box.

3 Plant the two verbenas at the back of the box between the geranium and ground ivies.

4 Plant the petunias between the geranium and ground ivies at the front of the box. Water thoroughly and stand in a sunny position.

GARDENER'S TIP

If the ground ivy gets too rampant, cut it back. Root some pieces in water and grow into plants to use elsewhere.

Plant in spring

Sweet-smelling Summer Flowers

Scented geranium and verbena are combined with heliotrope and petunias to make a window box that is a fragrant as well as a visual pleasure.

MATERIALS
40 cm (16 in) terracotta window box
Crocks or other suitable drainage material
Compost
Slow-release plant food granules

PLANTS
Scented-leaf geranium (*Pelargonium*) 'Lady Plymouth'
3 soft pink petunias
Heliotrope
2 *Verbena* 'Pink Parfait'

GARDENER'S TIP

At the end of the summer the geranium can be potted up and kept through the winter as a houseplant. Reduce the height of the plant by at least a half and it will soon send out new shoots.

Plant in late spring or early summer

1 Cover the base of the window box with a layer of crocks. Fill with compost, mixing in 2 teaspoons of slow-release plant food granules. Plant the geranium to the right of centre, towards the back of the window box.

2 Plant a petunia in each corner and one in the centre at the front of the window box.

3 Plant the heliotrope to the left of the geranium.

4 Plant one verbena behind the heliotrope and the other in front of the geranium. Water well and place in a sunny position.

Scented Geraniums

There is a wonderful variation in leaf size, shape and colouring as well as an incredible diversity of scents amongst the *Pelargonium* family. Choose the fragrances you like best and put the plants where you will brush against them to release their fragrance.

Materials
40 cm (16 in) terracotta window box
Crocks or other suitable drainage material
Compost
Slow-release plant food granules

Plants
4 scented-leaf geraniums (*Pelargonium fragrans*)

geraniums

1 Cover the base of the window box with a layer of crocks or other suitable drainage material. Fill with compost, mixing in 2 teaspoons of slow-release plant food granules. Plant the first geranium at the right-hand end of the container.

2 Choose a contrasting leaf colour and shape and plant this next to the first geranium towards the front edge of the window box.

3 Plant the third geranium behind the second geranium.

Gardener's Tip

During the summer, pick and dry the leaves of these geraniums for use in pot-pourri or in muslin bags to scent linen. If you have a greenhouse or conservatory, move the window box inside for the winter and water sparingly until spring.

Plant in spring

4 Finally, plant the fourth geranium at the left-hand end of the container. Water well and position in full or partial sun.

Sweet-scented Lavender

In this large basket an unusual lavender is planted amongst *Convolvulus sabatius* and the fan-shaped flowers of *Scaevola*. Underplanting ensures that the flowers cascade down the sides of the basket.

GARDENER'S TIP

If you are unable to obtain *Lavandula dentata*, then 'Hidcote' or 'Munstead' are easily available substitutes.

Plant in spring

MATERIALS
40 cm (16 in) hanging basket
Sphagnum moss
Compost
Slow-release plant food granules

PLANTS
3 *Convolvulus sabatius*
2 *Scaevola*
2 lavender (*Lavandula dentata* var. *candicans*)

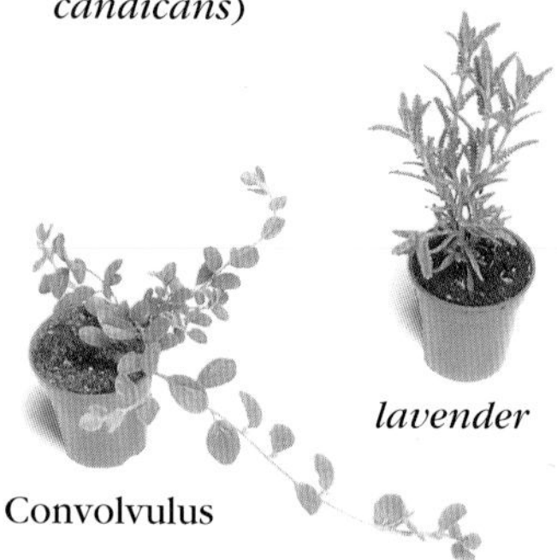

lavender

Convolvulus

Scaevola

1 Line the lower half of the basket with moss.

2 Plant two of the *Convolvulus* into the side of the basket by resting the rootballs on the moss and carefully guiding the foliage between the wires.

3 Plant one of the *Scaevola* into the side of the basket in the same way.

4 Line the rest of the basket with moss, taking care to tuck the moss around the underplanted plants.

5 Fill the basket with compost, mixing a teaspoon of slow-release plant food granules into the top layer. Plant the lavenders opposite one another in the top of the basket.

6 Plant the remaining *Convolvulus* and *Scaevola* plants in the spaces between the lavenders. Water thoroughly and hang in a sunny position.

Scented Window Box

The soft silvers and blues of the flowers and foliage beautifully complement this verdigris window box. The scent of the lavender and petunias will drift magically through open windows.

MATERIALS AND TOOLS
Window box, 60 cm (24 in) long
Gravel or similar drainage material
Equal mix loam-based compost and container compost
Slow-release plant food granules

PLANTS
2 lavender
2 pale blue petunias
3 deep blue petunias
3 *Chaenorrhinum glareosum* (lilac lobelia may be used instead)
6 *Helichrysum petiolare*

1 Fill the bottom 5 cm (2 in) of the window box with drainage material and then half-fill with a layer of compost. Position the lavender plants, loosening the soil around the roots before planting, as they will establish better this way.

2 Now arrange the flowering plants around the lavender, leaving spaces for the *Helichrysum* between them.

3 Finally add the foliage plants and fill between the plants with compost, pressing firmly so that no air gaps are left around the roots. Place in a sunny position and water regularly.

GARDENER'S TIP

To keep a densely planted container like this looking its best it is necessary to feed regularly with a liquid feed, or more simply to mix slow-release plant food granules onto the surface of the compost to last the whole summer. Cut back the lavender heads after flowering to ensure a bushy flowering plant again next year.

Plant in late spring or early summer.

A Nose-twitcher Window Box

One of the French country names for the nasturtium means 'nose-twitcher' and refers to the peppery smell of the plant. It has been planted here with the equally aromatic and colourful ginger mint and pot marigold.

MATERIALS
25 cm (10 in) terracotta window box
Crocks or other suitable drainage material
Compost
Slow-release plant food granules

PLANTS
Variegated ginger mint
Nasturtium 'Empress of India', or similar
Pot marigold (*Calendula*) 'Gitana', or similar compact form

1 Cover the base of the window box with a layer of crocks or similar drainage material. Fill the container with compost, mixing in a half-teaspoon of slow-release plant food granules.

2 Plant the ginger mint on the right of the container.

3 Plant the nasturtium in the centre.

4 Plant the marigold on the left of the container. Water well and stand in full or partial sun.

GARDENER'S TIP

A small window box like this one can double as a table centrepiece for an outdoor meal.

Plant in spring

A Butterfly Garden

We should all do our bit to encourage butterflies into our gardens and this window box with sedum, marjoram, thyme and origanum should prove irresistible. All these plants are perennials and can be over-wintered in the window box.

GARDENER'S TIP

You can imitate the look of an old stone window box by painting a new one with a dilute mixture of liquid seaweed plant food and water. This encourages moss to grow and "ages" the stone.

Plant in spring

MATERIALS
60 cm (24 in) stone window box
Crocks or other suitable drainage material
Compost
Slow-release plant food granules

PLANTS
Sedum 'Ruby Glow'
Marjoram
Lemon thyme (*Thymus citriodorus*)
Common thyme (*Thymus vulgaris*)
Origanum

1 Cover the base of the window box with a layer of crocks or other suitable drainage material. Fill with compost, mixing in 3 teaspoons of slow-release plant food granules.

2 Plant the sedum off-centre to the left of the window box.

3 Plant the marjoram to the left of the sedum.

4 Plant the lemon thyme in the centre at the front of the window box.

5 Plant the common thyme in the back right-hand corner of the container.

6 Plant the origanum in the front right-hand corner of the window box. Water well and place in a sunny position.

Star-jasmine in a Villandry Planter

The soft, seductive scent of the star-jasmine makes this a perfect container to be placed by the side of a door where the scent will be appreciated by all who pass through.

MATERIALS AND TOOLS
- Villandry planter or similar, approx 50 cm (18 in) square, preferably self-watering
- Equal mix loam-based compost and standard compost
- Slow-release plant food granules
- Bark mulch
- Trowel

PLANTS
- Star-jasmine (*Trachelospermum jasminoides*)

star-jasmine

1 Feed wicks through the holes in the base of the liner.

2 Fill the water reservoir in the base of the planter to the top of the overflow pipe and place the liner inside the planter.

3 Fill the bottom of the liner with compost while pulling through the wicks so that they reach the level of the jasmine roots.

4 Remove the jasmine from its pot, gently tease the roots loose and stand it in the planter.

5 Add compost and firm it around the rootball of the jasmine. Scatter 2 tablespoons of plant food granules on the surface and gently work it into the top layer of compost with the trowel.

6 Mulch around the plant with a layer of bark, then water. Check the reservoir of the self-watering container once a week and top up if necessary. Conventional pots should be watered daily in the early morning or evening during hot weather.

GARDENER'S TIP

Use a plastic liner inside all large planters. It is easier to remove the liner when replanting rather than to replant the entire container.

Plant in late spring or early summer.

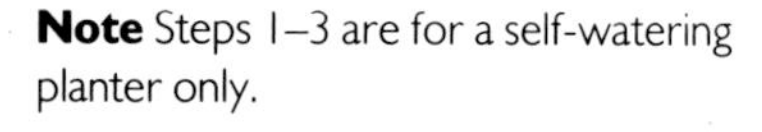

Note Steps 1–3 are for a self-watering planter only.

Shady Characters

All the plants used in this window box are perfectly happy in the shade. A periwinkle with variegated leaves and blue spring flowers is planted with blue-leaved hostas and summer-flowering busy lizzies in a window box that will brighten a gloomy corner for many months.

MATERIALS
45 cm (18 in) fibre window box
Crocks or other suitable drainage material
Compost
Slow-release plant food granules

PLANTS
Variegated periwinkle (*Vinca minor* 'Aureovariegata')
3 *Hosta* 'Blue Moon'
5 white busy lizzies (*Impatiens*)

periwinkle

Hosta

busy lizzie

1 Cover the base of the window box with a layer of drainage material. Fill the window box with compost, mixing in 2 teaspoons of slow-release plant food granules.

2 Plant the periwinkle in the centre of the box.

3 Plant two of the hostas in the back corners of the window box and the third in front of, or slightly to one side of, the periwinkle.

4 Plant the busy lizzies in the spaces between the other plants.

GARDENER'S TIP

To keep the busy lizzies looking their best, pick off the dead flowers and leaves regularly or they will stick to the plant and spoil its appearance.

Plant in late spring

Shade-loving Ferns

This window box is ideal for a dark, damp and shady spot - conditions that many plants dislike but much loved by ferns. Provided the plants are not allowed to dry out too often they will grow happily for many years.

MATERIALS
40 cm (16 in) glazed window box
Clay granules or other suitable drainage material
Compost
Slow-release plant food granules
Bark

PLANTS
A selection of three ferns

ferns

1 Cover the base of the container with a layer of drainage material.

2 Plant the first fern in the left-hand end of the container

3 Plant the second fern centrally.

4 Plant the third fern at the right-hand end of the window box. Mulch with bark and water thoroughly. Stand in a cool position.

GARDENER'S TIP

In the autumn when the leaves begin to die back, cut back all the foliage and apply a fresh layer of bark to protect the plants over winter. Feed in the spring.

Plant in spring

Shady Corner

Shady corners are often thought of as problematical, when in fact there is a wealth of wonderful plants that thrive in these situations, such as the hosta, hydrangea and fern used in this arrangement.

MATERIALS AND TOOLS
3 terracotta pots of various sizes
Crocks
Composted manure
Equal mix standard compost and loam-based compost
Trowel

PLANTS
Hosta sieboldiana elegans
Variegated hydrangea
Polystichum fern

Hosta sieboldiana elegans

variegated hydrangea

Polystichum *fern*

1 Plant the hosta in a pot large enough for its bulky root system and with space for further growth. The pot used here nicely echoes the shape of the leaves. Place crocks at the bottom of the pot and then a layer of manure before adding the potting compost. Follow this procedure with the hydrangea as well.

2 Plant the fern in a terracotta pot slightly larger than its existing pot. It should not need transplanting for two to three years.

3 The hydrangea makes a great deal of growth during the summer and will get very top heavy. Plant in a pot with plenty of space for root growth and heavy enough to prevent the plant toppling over.

GARDENER'S TIP

The hosta is a beautiful foliage plant much loved by slugs and snails which chew unsightly holes in the leaves. To prevent this, smear an inch-wide band of petroleum jelly below the rim of the container and the leaves will remain untouched.

Plant at any time of the year.

Full of Ferns

A damp shady corner is the perfect position for a basket of ferns. Provided they are regularly fed and watered, and the ferns are cut back in late autumn, this basket will give pleasure for many years. We have chosen hardy ferns, but the idea can be adapted for a conservatory or bathroom using less hardy plants such as the maidenhair fern.

MATERIALS
36 cm (14 in) hanging basket
Sphagnum moss
Compost
Slow-release plant food granules

PLANTS
4 different ferns (we used *Dryopteris*, *Athyrium*, *Matteuccia struthiopteris* and *Asplenium crispum*)

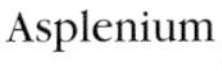

1 Line the basket with moss.

2 Fill the basket with compost. Mix a teaspoon of slow-release plant food granules into the top of the compost.

GARDENER'S TIP

Strange as it may seem, finely chopped banana skins are a favourite food of ferns. Simply sprinkle around the base of the stems and watch the ferns flourish.

Plant in spring

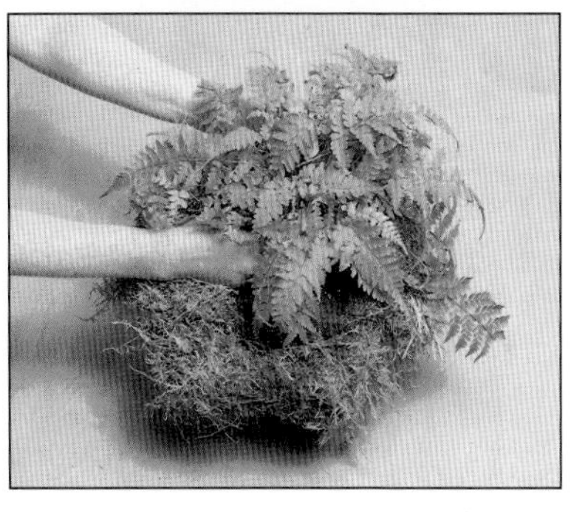

3 Before removing the ferns from their pots, arrange them in the basket to ensure that you achieve a balanced effect.

4 Plant the ferns.

A Dazzling Display

The succulents in this window box will provide a vivid splash of colour throughout the summer and are ideal for a hot dry windowsill. *Mesembryanthemums*, *Kalanchoë* and *Portulaca* all love the sunshine and will grow happily in this small window box.

MATERIALS
36 cm (14 in) plastic window box
Compost
Slow-release plant food granules

PLANTS
Kalanchoë
2 *Portulaca*
3 *Mesembryanthemum*

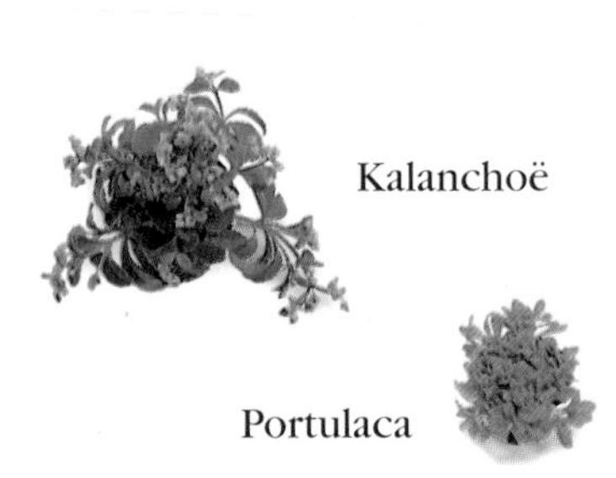

Kalanchoë

Portulaca

Mesembryanthemum

1 Check the drainage holes are open in the base and, if not, drill or punch them open. Fill the window box with compost, mixing in a teaspoon of slow-release plant food granules.

2 Plant the *Kalanchoë* in the centre of the window box.

3 Plant the two *Portulaca* in the front corners of the window box.

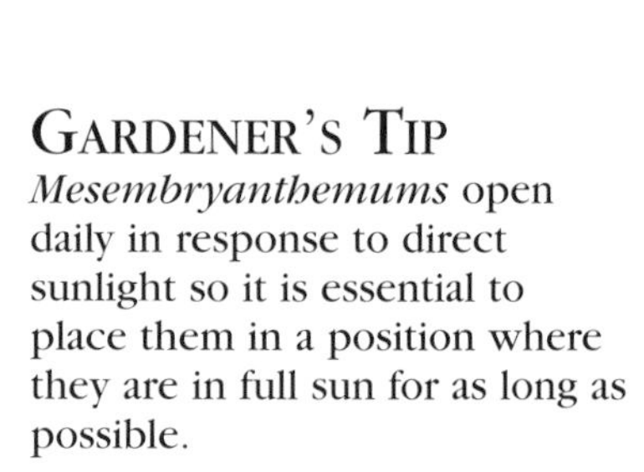

GARDENER'S TIP

Mesembryanthemums open daily in response to direct sunlight so it is essential to place them in a position where they are in full sun for as long as possible.

Plant in late spring or early summer

4 Plant one *Mesembryanthemum* in front of the *Kalanchoë* and the other two behind the two *Portulaca*. Water well and stand in a sunny position.

Desert Belles

An attractively weathered window box is the container used for this dramatic collection of succulents. With their architectural leaf shapes and wonderful range of colouring they would look particularly good in a contemporary setting.

Move the window box to a conservatory or frost-free greenhouse for the winter. Water sparingly only if plants show signs of shrivelling.

Plant in late spring or early summer for outdoor use, any time of year for a conservatory

MATERIALS
- 40 cm (16 in) terracotta window box
- Crocks or other suitable drainage material
- Compost
- Gravel
- Slow-release plant food granules

PLANTS
- Aloe
- *Crassula ovata*
- *Echeveria elegans*
- *Sansevieria trifasciata*

Aloe

Crassula ovata

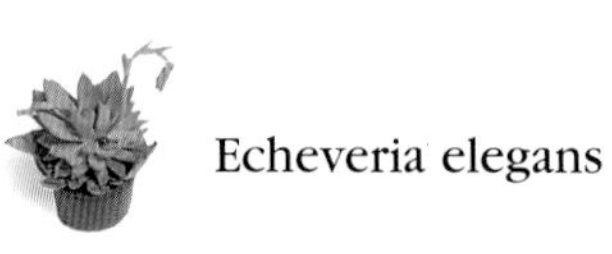

Echeveria elegans

Sansevieria trifasciata

1 Place a layer of crocks or other suitable drainage material in the base of the container.

2 Fill with compost, mixing in a teaspoon of slow-release plant food granules. Plant the aloe at the right-hand end of the window box.

3 Plant the *Crassula* next to the aloe.

4 Plant the *Echeveria* towards the back of the container, next to the *Crassula*.

5 Plant the *Sansevieria* in front of the *Echeveria*.

6 Surround the plants with a layer of gravel. Water well to establish and thereafter water sparingly. Place in full sun.

Bronze and Gold Winners

Bronze pansies and *Mimulus* and golden green *Lysimachia* take the medals in this striking arrangement, with richly coloured *Heuchera* adding to the unusual mixture of tones.

MATERIALS
40 cm (16 in) hanging basket
Sphagnum moss
Compost
Slow-release plant food granules

PLANTS
Heuchera 'Bressingham Bronze'
3 bronze-coloured pansies (*Viola*)
3 bronze-coloured *Mimulus*
3 *Lysimachia nummularia* 'Aurea'

Mimulus

pansy

Heuchera

Lysimachia

1 Line the basket with moss.

2 Fill the basket with compost, mixing a teaspoon of plant food granules into the top layer of compost.

3 Plant the *Heuchera* in the middle of the basket.

4 Plant the pansies, evenly spaced, around the *Heuchera*.

5 Plant the *Mimulus* between the pansies.

6 Plant the *Lysimachia* around the edge of the basket. Water well and hang in light shade.

GARDENER'S TIP

At the end of the season the *Heuchera* can be planted in the border or in a container. It will do best in partial shade, as full sun tends to scorch and discolour the leaves.

Plant in spring

A Spring Display of Auriculas

An old strawberry punnet carrier makes an attractive and unusual window box in which to display some beautifully marked auriculas planted in antique terracotta pots. A large flower basket or wooden trug would look just as good as this old wooden carrier.

MATERIALS
50 cm (20 in) wooden carrier
10 8-10 cm (3-4 in) old or antique-style terracotta pots
Crocks
Compost

PLANTS
10 different auriculas (*Primula auricula*)

auriculas

1 Place a crock over the drainage hole of a pot.

2 Remove a plant from its plastic pot and plant it firmly with added compost.

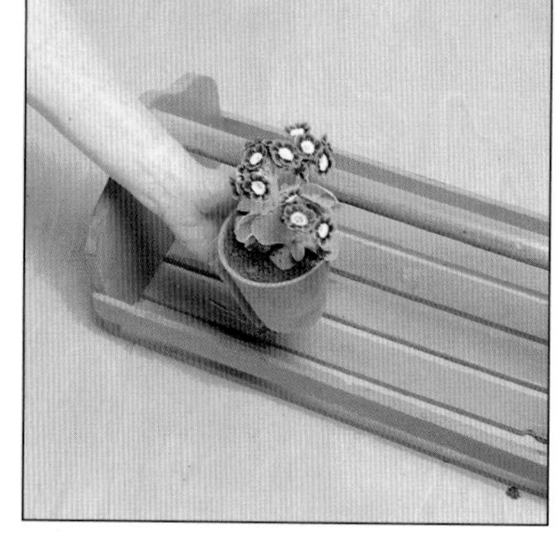

3 Stand the newly planted auricula in the wooden carrier.

4 Repeat the process for the other flowers. Water thoroughly and stand in light shade.

GARDENER'S TIP

A windowsill is an ideal position to see auriculas at their best. It is difficult to admire the full drama of their markings if they are at ground level. When they have finished flowering, stand the pots in a shady corner or a cold frame.

Plant in early spring

Spring Flowers

The dwarf narcissus 'Hawera' is surrounded with forget-me-nots and violas to create a delicately pretty spring display. While summer baskets need time to grow on in order to look their best, spring baskets give instant colour.

MATERIALS
30 cm (12 in) hanging basket
Sphagnum moss
Compost

PLANTS
Pot of dwarf narcissus 'Hawera', or similar
5 forget-me-not plants (*Myosotis*)
5 violas

violas

forget-me-not

narcissus

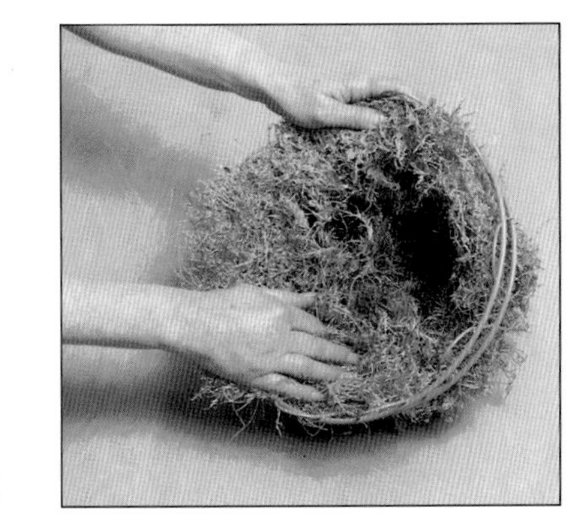

1 Line the basket with moss.

2 Remove the narcissus from the pot and place centrally in the basket. Fill the basket around the narcissus with compost.

3 Plant the forget-me-nots around the narcissus.

4 Plant the violas around the edge of the basket. Water well and hang in sun or shade.

GARDENER'S TIP

Keep the narcissus bulbs for next year by re-potting them when you dismantle the basket. Leave the foliage to die down naturally and they will flower again next year.

Plant in late winter or early spring

A Garland of Spring Flowers

Miniature daffodils, deep blue pansies, yellow polyanthus and variegated ivy are planted together to make a hanging basket that will flower for many weeks in early spring.

Plant in autumn if growing daffodils from bulb, and late winter or early spring for ready-grown daffodils

MATERIALS
30 cm (12 in) hanging basket
Sphagnum moss
Compost
Slow-release plant food granules

PLANTS
3 variegated ivies
5 miniature daffodil bulbs 'Tête-à-Tête', or similar
3 blue pansies (*Viola*)
2 yellow polyanthus

variegated ivy

polyanthus

pansy

miniature daffodils

1 Line the lower half of the basket with moss.

2 Plant the ivies into the side of the basket by resting the rootballs on the moss and guiding the foliage through the side of the basket.

3 Line the rest of the basket with moss and add a layer of compost to the bottom of the basket. Push the daffodil bulbs into the compost.

4 Fill the remainder of the basket with compost, mixing a teaspoon of slow-release plant food granules into the top layer. Plant the pansies in the top of the basket.

5 Plant the polyanthus between the pansies.

Spring Display in a Copper Boiler

A battered old wash boiler makes an attractive and characterful container for a display of white tulips underplanted with purple violets and evergreen periwinkles.

MATERIALS AND TOOLS
Copper boiler, 60 cm (24 in) diameter
Plastic pot, 20 cm (8 in) diameter
Standard compost
Trowel

PLANTS
20 white tulip bulbs or tulips in bud
5 purple violets
2 periwinkles (*Vinca minor*)

1 Place an upturned 20 cm (8 in) pot in the base of the boiler before filling it with compost. This will save on the amount of compost used and will not have any effect on the growth of the plants as they will still have plenty of room to grow.

2 If you are planting tulip bulbs, half-fill the container with compost, arrange the bulbs evenly over the surface and then cover them with a good 15 cm (6 in) of compost. This should be done in late autumn.

3 Do the underplanting in the early spring. The compost will have settled in the container and should be topped up to within 8 cm (3 in) of the rim. Remove the violets from their pots. Gently squeeze the rootballs and loosen the roots to aid the plants' growth.

4 Plant one violet in the centre and four around the edges. Scoop out the soil by hand to avoid damaging the growing tips of the tulips beneath the soil.

5 Plant a periwinkle either side of the central violet, again loosening the rootballs.

GARDENER'S TIP

Lift the tulips when they have finished flowering and hang them up to dry in a cool airy place. They can be replanted later in the winter to flower again next year. Provided you pick off the dead flowers the violets will flower all summer. For a summer display, lift the central violet and plant a standard white marguerite in the centre of the container.

Plant bulbs in autumn or plants in bud in spring. Plant the violets and periwinkle in spring.

6 Alternatively, if you are planting tulips in bud, the whole scheme should be planted at the same time. Work from one side of the pot to the other, interplanting the tulips with the violets and periwinkles. Press down firmly around the tulips or they will work themselves loose in windy weather. Position in sun or partial shade.

Terracotta Planter of Spring Bulbs

This container of bright yellow tulips and daffodils will brighten the dullest spring day. The variegated ivies conceal the soil and pleasantly soften the edge of the planter.

MATERIALS AND TOOLS
Terracotta planter, 60 cm (24 in) long
Crocks or similar drainage material
Standard compost
Trowel

PLANTS
10 tulips
6 pots of miniature daffodils
6 variegated ivies

1 Fill the bottom of the planter with drainage material. Be especially careful to cover the drainage holes so that they do not become clogged with compost.

2 Remove the tulips from their pots and carefully separate the bulbs. Plant them in a staggered double row down the length of the planter.

3 Interplant the container with the miniature daffodils.

4 Finally plant the ivies around the edge of the planter. Remember that if you are using a planter like this as a window box, the back of the arrangement should look just as good as the front.

GARDENER'S TIP

Plant in early spring.

Tinware Planter of Lily-of-the-valley

Lily-of-the-valley grow very well in containers and they will thrive in the shade where their delicate scented flowers stand out amongst the greenery. Surrounding the plants with bun moss is practical as well as attractive as it will stop the soil splashing back onto the leaves and flowers during spring showers.

MATERIALS AND TOOLS
Tinware planter
Clay granules
Standard compost
Bun moss
Trowel

PLANTS
6–8 pots of lily-of-the-valley

lily-of-the-valley

1 Fill the bottom of the planter with 5 cm (2 in) of clay granules.

2 Cover the granules with a layer of compost and place the lily-of-the-valley plants on the compost.

3 Fill in around the plants with more compost, making sure to press firmly around the plants so that they won't rock about in the wind. Now cover the soil with bun moss, fitting it snugly around the stems of the lily-of-the-valley, as the moss will also help keep the plants upright.

GARDENER'S TIP

If you want to bring your planter indoors to enjoy the scent of the flowers, use a planter without drainage holes in the base, but be very careful not to overwater. Once the plants have finished flowering replant them in a pot with normal drainage holes or in the garden. They are woodland plants and will be quite happy under trees.

Plant in early spring.

Miniature Spring Garden

Terracotta pots filled with crocuses, irises and primroses nestling in a bed of moss, make a delightful scaled-down spring garden which would fit on the smallest balcony or even a windowsill.

GARDENER'S TIP

After the primrose plants have finished flowering they will send out glossy green leaves all summer long if they are kept in a cool shady spot and are watered regularly. Next year, after flowering they can be divided up to provide you with many more plants.

MATERIALS AND TOOLS
Terracotta seed tray
2 terracotta pots, 12 cm (5 in) high
Crocks
Standard compost
Bun moss
Trowel

PLANTS
3 primroses
Pot of Reticulata irises in bud
Pot of crocuses in bud

crocus

Reticulata iris

bun moss

primrose

GARDENER'S TIP

Once the irises and crocuses are past their best, hide them behind other pots to die down and dry out before starting them into growth again in the autumn.

Plant in early spring.

1 Cover the drainage holes of the seed tray and the two pots with crocks.

2 Half-fill the seed tray with compost. Before planting the primroses, loosen the roots by gently squeezing the rootball and teasing the roots loose. The plants will establish themselves far better in the surrounding compost if you do this.

3 Arrange the primroses in the seed tray and, once you are happy with their positioning, fill in with compost around the plants, pressing down around the plants to ensure they are firmly planted.

4 Arrange the bun moss around the plants so that all the compost is hidden.

5 Remove the irises from their plastic pot and slip them into the terracotta pot. Bed them in with a little extra compost if necessary and then arrange moss around the base of the stems.

6 Repeat this process with the crocuses and then water all the pots.

Wooden Tub with Daffodils and Wallflowers

A weathered wooden tub planted in the autumn with daffodil bulbs and wallflower plants will provide a colourful spring display. Alternatively you can buy pots of daffodils and wallflowers in bud in the early spring for an instant display.

Materials and Tools
Wooden tub, 35 cm (14 in) diameter
Polystyrene plant tray
Standard compost
Slow-release plant food granules
Trowel

Plants
24 daffodil bulbs or 4 × 1 litre (5 in) pots of daffodils
3 bushy wallflower (*Cheiranthus*) plants

daffodils

wallflower

1 Break the polystyrene tray into large pieces and fill the bottom third of the tub to provide drainage and save on the quantity of compost used.

2 Add compost until the tub is half-full and arrange 12 of the daffodil bulbs evenly over the surface. Cover the bulbs with compost.

3 Arrange the other 12 bulbs on the surface of the compost.

Plant bulbs in the autumn or plants in bud in spring.

Gardener's Tip

To save the bulbs for next year allow the leaves to die right back and then dig up and store in a cool dry place.

4 Remove the wallflower plants from their pots and place them on the compost. Don't worry if the plants cover some of the bulbs, they will grow round the wallflowers. Fill the tub with compost, pressing down firmly around the wallflowers to ensure that they do not work loose in windy weather. Sprinkle a tablespoon of plant food granules onto the surface and work onto the top 3 cm (1 in) of compost.

Sapphires for Spring

Deep blue pansies are surrounded by gentian-blue *Anagallis* and underplanted with golden *Helichrysum* in this richly coloured basket.

MATERIALS
30 cm (12 in) hanging basket
Sphagnum moss
Compost
Slow-release plant food granules

PLANTS
3 *Helichrysum petiolare* 'Aureum'
3 deep blue pansies (*Viola*)
3 blue *Anagallis*

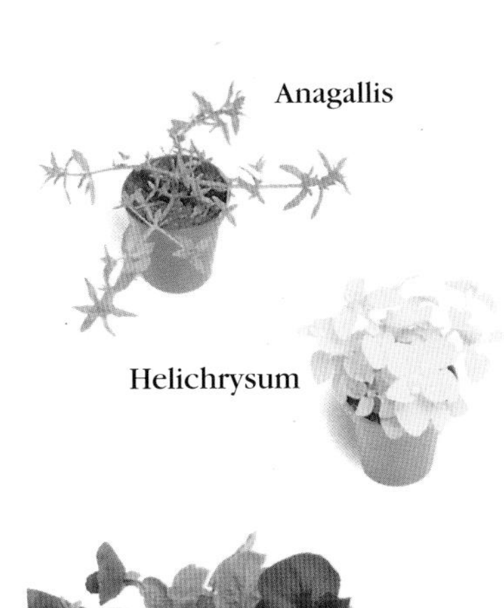

1 Line the lower half of the basket with moss.

2 Plant the *Helichrysum* in the sides of the basket by resting the rootballs on the moss and carefully guiding the foliage between the wires.

3 Line the rest of the basket with moss and fill with compost, mixing a teaspoon of slow-release plant food granules into the top layer. Plant the pansies, evenly spaced, in the top of the basket.

4 Plant the *Anagallis* between the pansies. Water thoroughly and hang in partial sun.

GARDENER'S TIP

The golden green colour of *Helichrysum petiolare* 'Aureum' is far stronger if the plants are not in full sun. Too much sun tends to fade the colouring.

Plant in spring

A Cottage Garden

Charming, cottage-garden plants tumble from this terracotta window box in a colourful display. The sunny flowers of the *Nemesia*, marigolds and nasturtiums mingle with the cool, soft green *Helichrysum* and blue-green nasturtium leaves.

MATERIALS
36 cm (14 in) terracotta window box
Crocks or other suitable drainage material
Compost
Slow-release plant food granules

PLANTS
3 pot marigolds (*Calendula*)
2 *Helichrysum petiolare* 'Aureum'
3 nasturtiums
2 *Nemesia* 'Orange Prince'

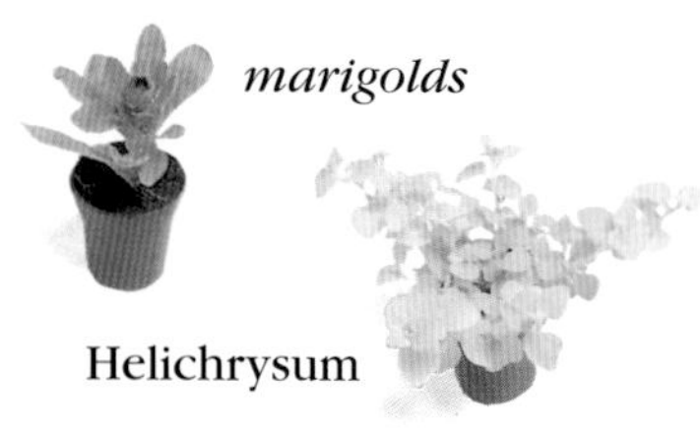

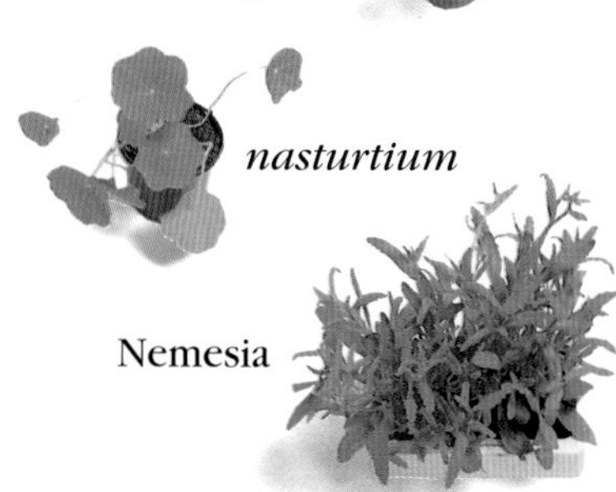

GARDENER'S TIP

The golden-leaved *Helichrysum* retains a better colour if it is not in full sun all day. Too much sun and it looks rather bleached.

Plant in spring

1 Cover the base of the container with crocks or similar suitable drainage material. Fill the window box with compost, mixing in 2 teaspoons of slow-release plant food granules. Plant the marigolds, evenly spaced along the back of the container.

2 Plant the two *Helichrysum* in the front corners of the window box.

3 Plant the nasturtiums between the marigolds at the back of the container.

4 Plant the *Nemesia* between the *Helichrysum* at the front of the window box. Water well and stand in partial sun.

Layers of Flowers

This window box is unusual as the colours are in distinct layers, with upright white flowering tobacco above pink impatiens and tumbling white variegated geranium and lobelias. The fibre window box is concealed by a decorative twig container.

MATERIALS
36 cm (14 in) fibre window box
Drainage material
Compost
Slow-release plant food granules

PLANTS
2 white flowering tobacco (*Nicotiana*)
Variegated geranium (*Pelargonium*) 'l'Elégante'
2 pink busy lizzies (*Impatiens*)
3 white lobelia

GARDENER'S TIP

Somehow, rogue blue lobelias have appeared amongst the white plants. This sort of thing often happens in gardening and, as in this case, the accidental addition often works well.

Plant in spring

1 Put a layer of drainage material in the base of the window box; fill with compost, mixing in 2 teaspoons of slow-release plant food granules. Plant the flowering tobacco on either side of the centre, near the back edge.

2 Plant the geranium at the front of the window box, in the centre.

3 Plant the busy lizzies at either end of the window box.

4 Plant one of the lobelias between the flowering tobacco and the other two on either side of the geranium.

Sweet Peas, Geranium and Chives

This large basket is filled with sweet peas surrounding a regal geranium (*Pelargonium*) and interplanted with chives to provide a contrasting leaf shape and help deter pests.

GARDENER'S TIP

Sweet peas will flower longer if you keep picking the flowers and be sure to remove any seed pods as they form. Similarly, the chives grow longer and stronger if their flower heads are removed before they seed.

Plant in late spring

MATERIALS
40 cm (16 in) hanging basket
Sphagnum moss
Compost
Slow-release plant food granules

PLANTS
Regal geranium (*Pelargonium*) 'Sancho Panza'
2-3 small pots or a strip of low-growing sweet peas such as 'Snoopea'
3 chive plants

sweet peas

chives

geranium

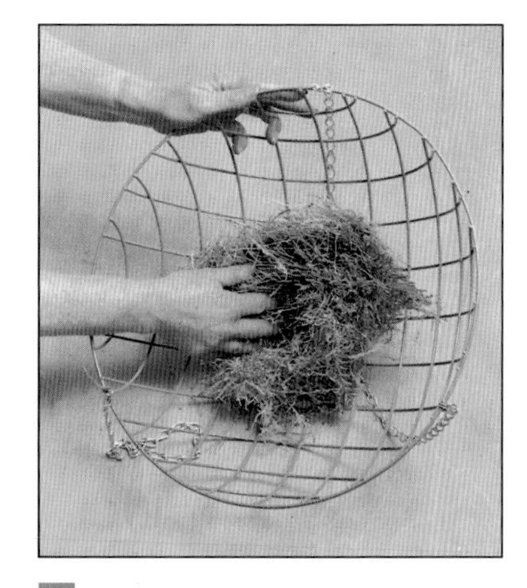

1 Line the basket with moss.

2 Fill the basket with compost and mix a teaspoon of slow-release plant food granules into the top of the compost. Plant the geranium (*Pelargonium*) in the centre of the basket.

3 Gently divide the sweet peas into clumps of about eight plants each.

4 Plant the sweet pea clumps around the edge of the basket.

5 Plant the chives between the sweet peas and the geranium.

6 Fill any gaps with a little moss. Water well and hang in a sunny position.

Marguerites and Pimpernels

We are more familiar with the wild scarlet pimpernel, but in this window box its blue relative, *Anagallis*, has been planted to climb amongst the stems of the yellow marguerites and snapdragons. Blue-flowered variegated *Felicia* and golden *Helichrysum* complete the picture.

GARDENER'S TIP

Dead-head the marguerites, snapdragons and *Felicia* to keep them flowering all summer. When planting the marguerites, pinch out the growing tips to encourage bushy plants.

Plant in spring

MATERIALS
76 cm (30 in) plastic window box
Compost
Slow-release plant food granules

PLANTS
2 yellow marguerites (*Argyranthemum*) 'Jamaica Primrose'
4 blue *Anagallis*
3 variegated *Felicia*
3 *Helichrysum petiolare* 'Aureum'
4 yellow snapdragons (*Antirrhinum*)

1 Check the drainage holes are open in the base and, if not, drill or punch them open. Fill the window box with compost, mixing in 3 teaspoons of slow-release plant food granules.

2 Plant the marguerites on either side of the centre in the middle of the window box.

3 Plant two of the *Anagallis* in the back corners of the window box and the other two at the front, on either side of the marguerites.

4 Plant one *Felicia* in the centre of the box and the other two on either side of the *Anagallis*.

5 Plant the *Helichrysum* in the front corners of the window box.

6 Plant two of the snapdragons on either side of the central *Felicia* and the other two on either side of the marguerites. Water thoroughly, drain, and stand in a sunny or partially sunny position.

Summer Pansies with Daisies and Convolvulus

Pale orange pansies contrast beautifully with the lavender-blue *Convolvulus* and the pastel yellow *Brachycome* daisies link the whole scheme together.

MATERIALS
30 cm (12 in) hanging basket
Sphagnum moss
Compost
Slow-release plant food granules

Plants
3 orange pansies (*Viola*)
3 *Brachycome* 'Lemon Mist'
2 *Convolvulus sabatius*

GARDENER'S TIP

Each time you water this basket be sure to remove any pansy flowers that are past their best. Once pansies start to set seed they quickly get "leggy" and stop flowering.

Plant in spring

1 Line the basket with moss.

2 Fill the basket with compost, mixing a teaspoon of slow-release plant food granules into the top layer. Plant the pansies, evenly spaced in the top of the basket.

3 Plant the *Brachycome* daisies between the pansies.

4 Plant the *Convolvulus* plants in the centre of the basket so that the tendrils can weave between the other plants. Water and hang in full or partial sun.

A Lime-green and Blue Window Box

Lime-green flowering tobacco and *Helichrysum* contrast beautifully with the blue *Scaevola* and *Convolvulus* in this window box of cool colours.

MATERIALS
76 cm (30 in) plastic window box
Compost
Slow-release plant food granules

PLANTS
5 lime-green flowering tobacco (*Nicotiana*)
2 *Scaevola*
2 *Helichrysum petiolare* 'Aureum'
3 *Convolvulus sabatius*

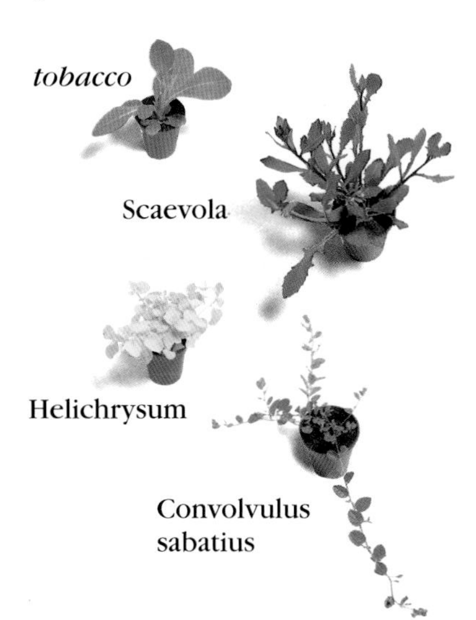

GARDENER'S TIP

At the end of the season you can pot up the *Scaevola* and *Convolvulus* plants to use again next year. Cut right back and overwinter on a windowsill or in a frost-free greenhouse.

Plant in late spring or early summer

1 Check the drainage holes are open in the base and, if not, drill or punch them open. Fill the window box with compost, mixing in 3 teaspoons of slow-release plant food granules. Plant the tobacco plants evenly spaced at the rear of the window box.

2 Plant the two *Scaevola* plants approximately 10 cm (4 in) from each end, in front of the tobacco plants.

3 Plant the two *Helichrysum* plants on either side of the centre of the window box next to the *Scaevola*.

4 Plant two of the *Convolvulus* in the front corners of the box and the third in the centre, at the front. Water thoroughly and position in light shade or partial sun.

A Pastel Composition

Pure white geraniums (*Pelargoniums*) emerge from a sea of blue *Felicia*, pinky-blue *Brachycome* daisies and verbena in this romantic basket.

MATERIALS
36 cm (14 in) hanging basket
Sphagnum moss
Compost
Slow-release plant food granules

PLANTS
2 pink verbena
2 *Brachycome* 'Pink Mist'
Blue *Felicia*
White geranium (*Pelargonium*)

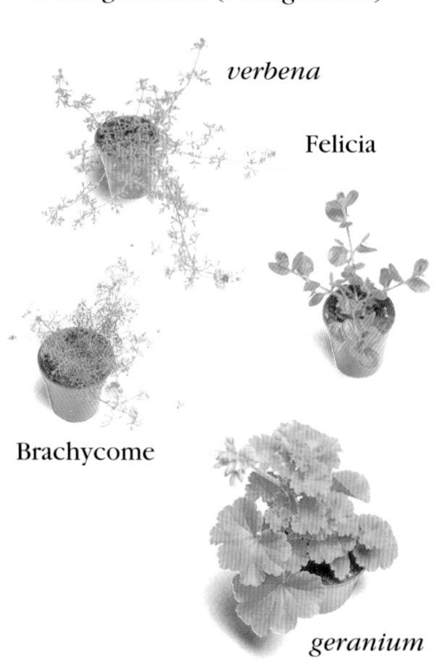

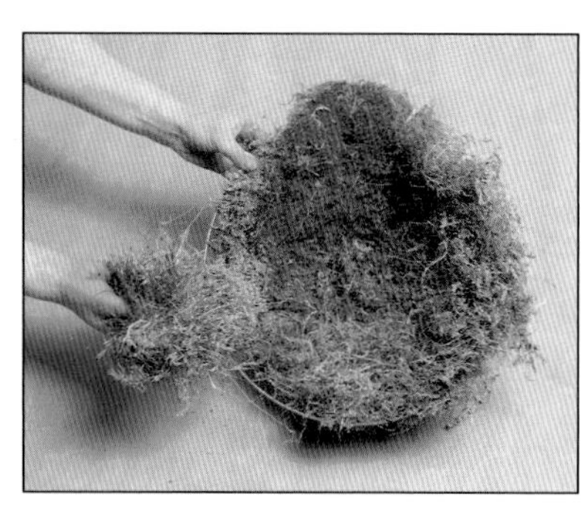

1 Line the basket with moss.

2 Fill the basket with compost, mixing a teaspoon of slow-release plant food granules into the top layer.

3 Plant the verbenas opposite each other at the edge of the basket, angling the rootballs to encourage the foliage to tumble over the sides.

4 Plant the *Brachycome* daisies around the edge of the basket.

GARDENER'S TIP

White geranium flowers discolour as they age; be sure to pick them off to keep the basket looking its best.

Plant in late spring or early summer

5 Plant the *Felicia* off-centre in the middle of the basket.

6 Plant the geranium (*Pelargonium*) off-centre in the remaining space in the middle of the basket. Water thoroughly and hang in a sunny position.

An Antique Wall Basket

This old wirework basket is an attractive container for a planting scheme which includes deep pink pansies, a variegated-leaf geranium (*Pelargonium*) with soft pink flowers, a blue *Convolvulus* and deep pink alyssum.

GARDENER'S TIP

Wall baskets look good amongst climbing plants, but you will need to cut and tie back the surrounding foliage if it gets too exuberant.

Plant in late spring or early summer

MATERIALS
30 cm (12 in) wall basket
Sphagnum moss
Compost
Slow-release plant food granules

PLANTS
5 rose-pink alyssum
Ivy-leaved geranium (*Pelargonium*) 'L'Elégante'
3 deep pink pansies
Convolvulus sabatius

alyssum

geranium

pansy

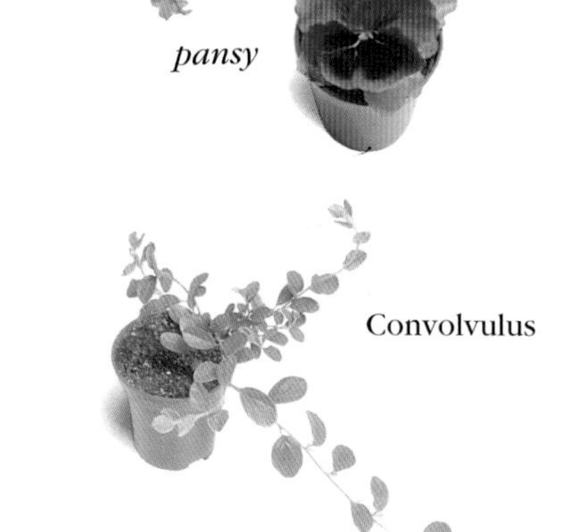

Convolvulus

1 Line the back of the basket and the lower half of the front with moss. Plant the alyssum into the side of the basket by resting the rootballs on the moss and guiding the foliage through the wires.

2 Line the remainder of the basket with moss and fill with compost, mixing a half-teaspoon of plant food granules into the top layer of compost. Plant the geranium at the front of the basket.

3 Plant the pansies around the geranium.

4 Plant the *Convolvulus* at the back of the basket, trailing its foliage through the other plants. Water well and hang in partial sun.

Daisy Chains

The soft yellows of the marguerite flowers and foliage are emphasized by combining them with bright blue *Felicia* flowers in this summery basket.

MATERIALS
40 cm (16 in) hanging basket
Sphagnum moss
Compost
Slow-release plant food granules

PLANTS
3 variegated *Felicia*
3 yellow marguerites (*Argyranthemum*)
3 *Helichrysum petiolare* 'Aureum'

1 Line the lower half of the basket with moss.

2 Plant the *Felicia* into the side of the basket by resting the rootballs on the moss and carefully guiding the foliage through the sides of the basket.

3 Line the rest of the basket with moss. Fill the basket with compost, mixing a teaspoon of slow-release plant food granules into the top layer. Plant the marguerites in the top of the basket.

4 Plant the *Helichrysum* between the marguerites, angling the plants to encourage them to grow over the edge of the basket. Water well and hang in full or partial sun.

GARDENER'S TIP

Pinch out the growing tips of the marguerites regularly to encourage bushy plants.

Plant in late spring or early summer

A Trough of Alpines

A selection of easy-to-grow alpine plants are grouped in a basket-weave stone planter to create a miniature garden. The mulch of gravel is both attractive and practical as it prevents soil splashing on to the leaves of the plants.

GARDENER'S TIP

Tidy the trough once a month, removing dead flower heads and leaves and adding more gravel if necessary. A trough like this will last a number of years before it needs replanting.

Plant in spring

MATERIALS

40 cm (16 in) stone trough
Crocks or other suitable drainage material
Compost
Slow-release plant food granules
Gravel

PLANTS

Sempervivum
Alpine *Aquilegia*
White rock rose (*Helianthemum*)
Papaver alpinum
Alpine phlox
Pink saxifrage
White saxifrage

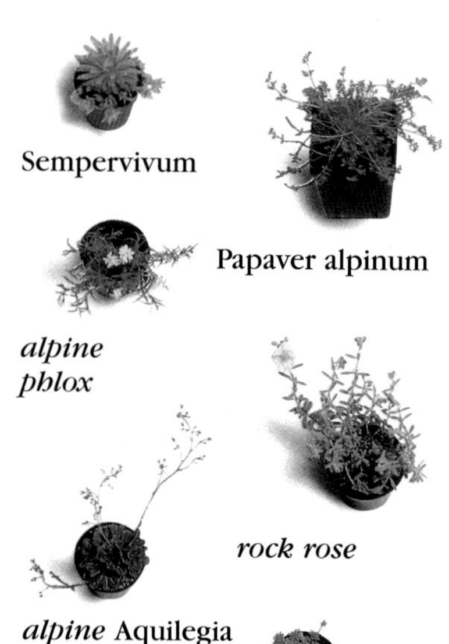

1 Cover the base of the trough with a layer of crocks.

2 Fill the container with compost, mixing in a teaspoon of slow-release plant food granules and extra gravel for improved drainage.

3 Before planting, arrange the plants, still in their pots, in the trough to decide on the most attractive arrangement.

4 Start planting from one end, working across the trough.

5 Complete the planting.

6 Scatter a good layer of gravel around the plants. Water thoroughly and stand in a sunny position.

Old Favourites

Dianthus, violas and candytuft are delightful cottage-garden plants and make a pretty display during late spring and early summer. Although by the time we took our photograph the candytuft had finished flowering the other flowers were still putting on a good show.

Materials
40 cm (16 in) painted wooden window box
Crocks or other suitable drainage material
Compost
Slow-release plant food granules

Plants
Dianthus
Candytuft (*Iberis*)
2 violas

Dianthus

candytuft

viola

1 Cover the base of the container with a layer of drainage material. Fill with compost, mixing in a teaspoon of slow-release plant food granules.

2 Plant the *Dianthus* slightly to the right of the centre of the window box.

3 Plant the candytuft to the left of the centre of the window box.

4 Plant a viola at each end. Water well and stand in a mixture of sun and shade.

Gardener's Tip

Once the flowers are over, cut the plants back and plant them out in the garden. There is still time to replant the window box with summer plants.

Plant in early spring

Sea View

While all these plants are most at home in a seaside environment, they are also happy on a hot windowsill away from the coast. Scatter a few seashells around them and you have got your own private beach!

Materials
45 cm (18 in) terracotta tray
4 10 cm (4 in) terracotta pots
Crocks
Compost
Slow-release plant food granules
Gravel
Seashells

Plants
Orange *Gazania*
2 yellow *Osteospermum*
Yellow *Portulaca*

Gardener's Tip

At the end of the summer the *Gazania* and *Osteospermums* can be overwintered by cutting them back and keeping them fairly dry in a frost-free environment. Next spring they can be planted out in the garden.

Plant in late spring or early summer

1 Plant the *Gazania* in a terracotta pot. When potting in terracotta, place a crock over the drainage hole in the base of the pot.

2 Plant each *Osteospermum* in a terracotta pot.

3 Plant the *Portulaca* in a terracotta pot. When all the plants are re-potted, divide a teaspoon of plant food granules between the four pots, working them into the top layer of compost.

4 Cover the tray with a layer of gravel.

5 Arrange the plants on the tray.

6 Mulch the pots with gravel and add seashells to the pots and the tray. Water well and stand in a sunny position.

Daring Reds and Bold Purples

The colour of the fuchsia flowers is echoed by the deep purple and crimson petunias in this window box, which also includes trailing campanula and catmint.

GARDENER'S TIP

At the end of the season the catmint plants can be trimmed back and planted in the garden. The fuchsia and *Campanulas* can be cut back and potted up to be overwintered in a frost-free greenhouse.

Plant in spring

MATERIALS
76 cm (30 in) plastic window box
90 cm (3 ft) wooden window box (optional)
Compost
Slow-release plant food granules

PLANTS
Fuchsia 'Dollar Princess'
2 low-growing catmint (*Nepeta mussinii*)
2 white-flowered *Campanula isophylla*
2 crimson petunias
2 purple petunias

1 Check the drainage holes are open in the base and, if not, drill or punch them open. Fill the window box with compost, mixing in 3 teaspoons of slow-release plant food granules. Plant the fuchsia in the centre of the window box.

2 Plant the catmint at either end of the window box.

3 Plant the *Campanula* next to the catmint.

4 Plant the crimson petunias on either side of the fuchsia at the back of the window box.

5 Plant the purple petunias on either side of the fuchsia at the front of the window box. Water thoroughly and allow to drain.

6 Lower the plastic window box into place inside the wooden window box, if using. Stand in a sunny position.

Filigree Foliage

The purply-black leaves of the *Heuchera* are all the more stunning when surrounded by the delicate silver-and-green filigree foliage of *Senecio*, the tender lavender *pinnata* and the soft lilac-coloured flowers of the *Bacopa* and the *Brachycome* daisies. The plants are grown in a white plastic planter which is concealed by an elegant wooden window box.

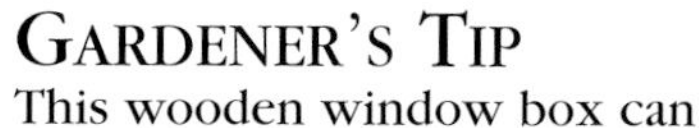

GARDENER'S TIP

This wooden window box can be set up to be self-watering and is ideal where access is difficult for daily watering. A variety of self-watering containers are available and come with full instructions on how they work.

Plant in spring

MATERIALS
76 cm (30 in) plastic window box
Compost
Slow-release plant food granules
90 cm (3 ft) wooden window box (optional)

PLANTS
Heuchera 'Palace Purple'
2 *Lavandula pinnata*
2 blue *Brachycome* daisies
3 *Senecio Cineraria* 'Silver Dust'
2 blue *Bacopa*

1 Check drainage holes are open in base of planter and, if not, drill or punch them out. Fill the window box with compost, mixing in 2 teaspoons of slow-release plant food granules. Plant the *Heuchera* in the centre.

2 Plant the two lavenders on either side of the *Heuchera*.

3 Plant the two *Brachycome* daisies at either end of the window box.

4 Plant the three *Senecio* at the front of the box next to the *Brachycome*.

5 Plant the two *Bacopa* between the *Senecio* and the *Heuchera*.

6 Water thoroughly and lift into place in the wooden window box, if using. Place in full or partial sun.

Mediterranean Garden

The brilliant colours of the Mediterranean are re-created with these painted pots. The plants thrive in the climate of the Mediterranean, but will also perform well in less predictable weather.

MATERIALS AND TOOLS
4 terracotta pots of various sizes
Paintbrush
Selection of brightly coloured emulsion paints
Masking tape
Crocks
Loam-based compost with ⅓ added grit
Gravel

PLANTS
Prostrate rosemary
Aloe
Golden thyme
Large red geranium

geranium

aloe

prostrate rosemary

golden thyme

1 Paint the pots with solid colours or with patterns. The paints used here are thicker than ordinary emulsion, so you may need two coats to get the same effect. The terracotta absorbs the moisture from the paint, so they will dry very quickly.

2 Paint the rim of one pot with a contrasting colour.

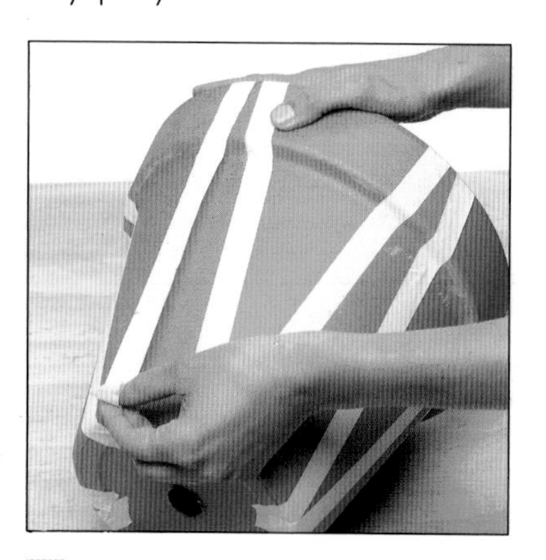

3 Create a pattern using tape to mask out specific areas.

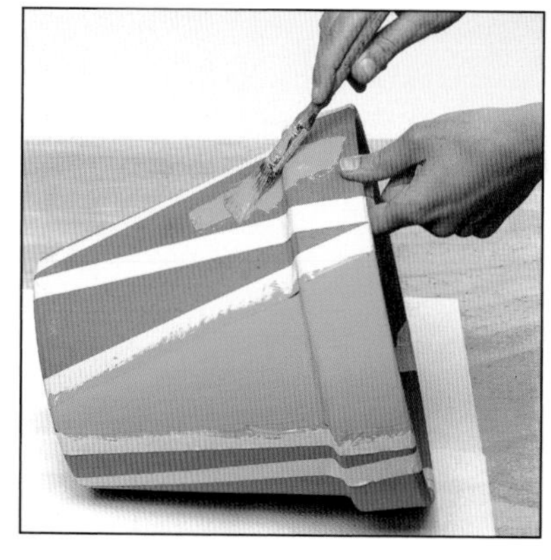

4 Paint every other area to create a zig-zag effect.

5 Place crocks in the bottom of the pots and then position the plants, firming them in place with extra compost. The roots of this rosemary are compacted and will benefit from being teased loose before planting.

GARDENER'S TIP

For commercial reasons the plants you buy will probably have been grown in a peat compost, although they prefer a loam-based compost. Gently loosen the peat around their roots and mix it with the loam-based compost before potting them up in the new mixture.

Plant in late spring or early summer.

6 The aloe does not need a large pot. Plant it in a pot just slightly larger than the one you bought it in.

7 Plant the thyme and geranium in separate pots. Finish the plants with a top-dressing of gravel, water well and place in a sheltered sunny corner.

A Space in the Sun

Osteospermum, *Portulaca* and *Diascia* are all sun-lovers so this is definitely a basket for your sunniest spot where the plants will thrive and the colours will look their best.

MATERIALS
36 cm (14 in) hanging basket
Sphagnum moss
Compost
Slow-release plant food granules

PLANTS
6 peach *Portulaca*
Osteospermum 'Buttermilk'
3 *Diascia* 'Salmon Supreme', or similar

Portulaca

Diascia

Osteospermum

1 Line the lower half of the basket with moss.

2 Plant three of the *Portulaca* into the side of the basket by resting the rootball on the moss and carefully guiding the foliage between the wires.

3 Add more moss to the basket, tucking the moss carefully around the *Portulaca* plants.

4 Partly fill the basket with compost, mixing a teaspoon of slow-release plant food granules into the top layer. Plant the remaining three *Portulaca* into the side of the basket, just below the rim.

5 Complete lining the basket with moss. Plant the *Osteospermum* centrally.

GARDENER'S TIP

Keep pinching out the growing tips of the *Osteospermum* to ensure a bushy plant.

Plant in late spring or early summer

6 Plant the *Diascia* around the *Osteospermum*. Water thoroughly and hang in a sunny spot.

Summer Carnival

The orange markings on the throats of some of the *Mimulus* flowers look wonderful with the orange-flowered geranium (*Pelargonium*) in this colourful basket. By the end of the season, trails of *Lysimachia* leaves will form a waterfall of foliage round the base.

MATERIALS
36 cm (14 in) basket
Sphagnum moss
Compost
Slow-release plant food granules

PLANTS
Orange-flowered geranium (zonal *Pelargonium*)
3 *Lysimachia nummularia* 'Aurea'
3 *Mimulus*

Mimulus
Lysimachia
geranium

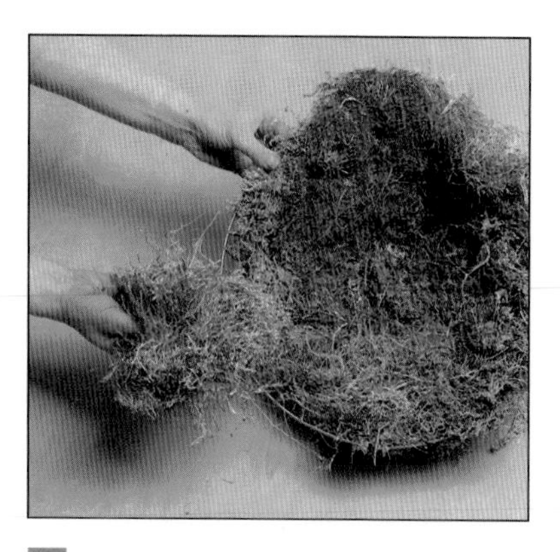

1 Line the basket with moss.

2 Fill the basket with compost, mixing a teaspoon of slow-release plant food granules into the top layer. Plant the geranium (*Pelargonium*) in the centre of the basket.

3 Plant the *Lysimachia*, evenly spaced around the edge of the basket, angling the plants to encourage them to trail over the sides.

4 Plant the *Mimulus* between the *Lysimachia*. Water thoroughly and hang in a sunny spot.

GARDENER'S TIP

Dead-head the flowers regularly to encourage repeat flowering and if the *Mimulus* start to get "leggy" cut back the offending stems to a leaf joint. New shoots will soon appear.

Plant in late spring or early summer

Fire and Earth

The earth tones of this small decorative terracotta window box are topped with the fiery reds and oranges of the plants – the fuchsia with its bronze foliage and tubular scarlet flowers, the orange nasturtiums and the red claw-like flowers of the feathery-leaved lotus.

MATERIALS
36 cm (14 in) terracotta window box
Clay granules or similar drainage material
Compost
Slow-release plant food granules

PLANTS
Fuchsia fulgens 'Thalia'
3 orange nasturtiums 'Empress of India', or similar
2 *Lotus berthelotti*

GARDENER'S TIP

This stunning fuchsia is worth keeping for next year. Pot it up in the autumn, cut back by half and overwinter on a windowsill or in a heated greenhouse.

Plant in late spring or early summer

1 Cover the base of the window box with drainage material.

2 Fill the window box with compost, mixing in a teaspoon of slow-release plant food granules. Plant the fuchsia in the centre of the window box.

3 Plant the three nasturtiums, evenly spaced, along the back of the window box.

4 Plant the two lotuses in the front of the window box on either side of the fuchsia. Water thoroughly, leave to drain, and stand in a sunny position.

A Wall Basket of Contrasting Colours

The deep green and burgundy foliage of *Fuchsia* 'Thalia' will be even more startling later in summer when the bright red pendant flowers stand out against the leaves and compete with the glowing colours of the *Nemesia*. The yellow-green *Helichrysum* provides a cooling contrast.

MATERIALS
30 cm (12 in) wall basket
Sphagnum moss
Compost
Slow-release plant food granules

PLANTS
3 *Helichrysum petiolare* 'Aureum'
Fuchsia 'Thalia'
4 *Nemesia* in red, yellow and orange tones

Helichrysum

Nemesia

Fuchsia

GARDENER'S TIP

Dead-head the *Nemesia* regularly to ensure that they continue flowering throughout the summer.

Plant in late spring or early summer

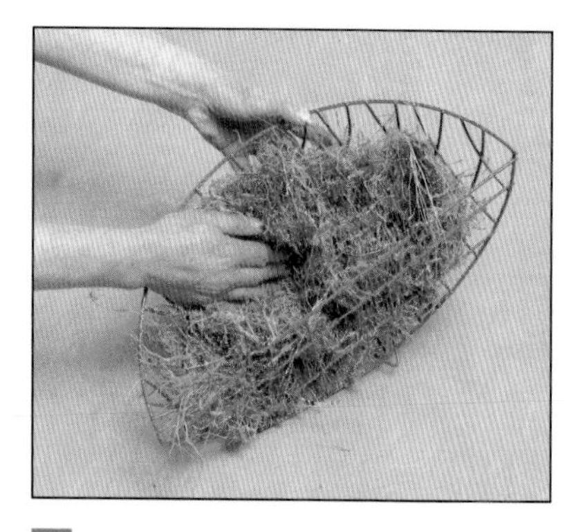

1 Line the back of the basket and the lower half of the front with moss. Fill the lower half of the basket with compost.

3 Line the rest of the basket with moss and top up with compost. Mix a half-teaspoon of slow-release plant food granules into the top layer of compost. Plant the fuchsia in the centre of the basket.

2 Plant two of the *Helichrysum* plants into the side of the basket by resting the rootballs on the moss and carefully feeding the foliage between the wires.

4 Plant the remaining *Helichrysum* in front of the fuchsia. Plant two *Nemesia* on either side of the central plants. Water well and hang in full or partial sun.

Small is Beautiful

Not everyone has room for a large hanging basket, especially when the plants have reached maturity, but there is sure to be space for a small basket like this one which will flower cheerfully all summer long.

MATERIALS
25 cm (10 in) hanging basket
Sphagnum moss
Compost
Slow- release plant food granules

PLANTS
4 nasturtiums
2 *Lysimachia nummularia* 'Aurea'
3 pot marigolds (*Calendula*)

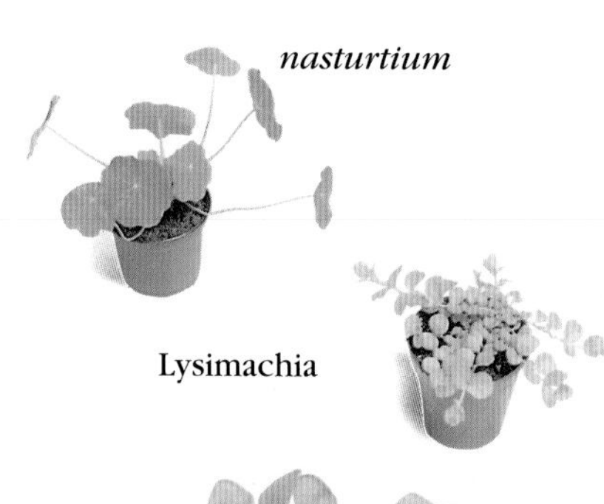

nasturtium

Lysimachia

marigolds

1 Line the lower half of the basket with moss.

2 Plant three of the nasturtiums into the side of the basket by resting the rootballs on the moss and carefully guiding the leaves through the sides of the basket.

3 Line the rest of the basket with moss and fill with compost, mixing a half-teaspoon of slow-release plant food granules into the top layer. Plant the *Lysimachia* opposite one another at the edge of the basket.

4 Plant the pot marigolds in the top of the basket.

5 Plant the remaining nasturtium in the middle of the basket. Water well and hang in a sunny position.

GARDENER'S TIP

Small baskets dry out very quickly so be sure to water frequently. To give a really good soak, you can immerse the basket in a bucket of water, but be careful not to damage the trailing plants.

Plant in spring

Dark Drama

The intense purple of the heliotrope usually dominates other plants, but here it is teamed with a selection of equally dramatic plants - *Dahlia* 'Bednall Beauty', with its purple foliage and dark red flowers, black grass and red and purple verbenas - in a stunning display.

MATERIALS
60 cm (24 in) terracotta window box
Broken polystyrene or other suitable drainage material
Compost
Slow-release plant food granules

PLANTS
Heliotrope
2 *Dahlia* 'Bednall Beauty'
Black grass (*Ophiopogon*)
2 purple trailing verbenas
2 red trailing verbenas

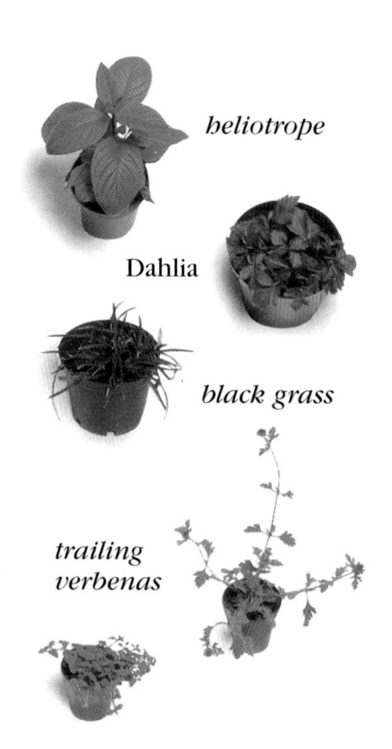

1 Fill the bottom of the window box with suitable drainage material.

2 Fill the window box with compost, mixing in 3 teaspoons of slow-release plant food granules. Plant the heliotrope centrally at the back of the window box, gently teasing loose the roots, if necessary.

3 Plant the dahlias in the back corners of the window box.

4 Plant the black grass in front of the heliotrope.

5 Plant the purple verbenas at the back between the heliotrope and the dahlias.

GARDENER'S TIP

Dahlias can be overwintered by digging up the tubers after the first frosts, cutting the stems back to 15 cm (6 in) and drying them off before storing in slightly damp peat in a frost-free shed. Start into growth again in spring and plant out after all danger of frost is past.

Plant in late spring or early summer

6 Plant the red verbenas at the front in either corner. This is a large container so it is best to position it before watering. Put it where it will benefit from full sun, then water thoroughly.

Divine Magenta

The gloriously strong colour of magenta petunias is combined with blue *Convolvulus sabatius*, heliotrope, which will bear scented deep purple flowers, and a variegated scented-leaf geranium (*Pelargonium*), which will add colour and fragrance later in the summer.

GARDENER'S TIP

Baskets with flat bases like this one can be stood on columns rather than hung from brackets. This is a useful solution if fixing a bracket is difficult.

Plant in late spring or early summer

MATERIALS
45 cm (18 in) basket
Sphagnum moss
Compost
Slow-release plant food granules

PLANTS
Scented-leaf geranium (*Pelargonium*) 'Fragrans Variegatum'
3 purple heliotropes
3 *Convolvulus sabatius*
5 trailing magenta-flowered petunias

1 Line the basket with moss.

2 Fill the basket with compost, mixing a teaspoon of slow-release plant food granules into the top layer.

3 Plant the scented-leaf geranium (*Pelargonium*) in the middle of the hanging basket.

4 Plant the heliotropes, evenly spaced around the geranium.

5 Plant the *Convolvulus*, evenly spaced around the edge of the basket.

6 Plant the petunias between the *Convolvulus* and the heliotrope.

Sugar and Spice

The candy-floss colour of the petunias is enriched by combining them with deep crimson ivy-leaved geraniums (*Pelargonium*). Slower growing silver-leaved snapdragons and a variegated geranium will add further colour later in the summer.

MATERIALS
36 cm (14 in) hanging basket
Sphagnum moss
Compost
Slow-release plant food granules

PLANTS
3 snapdragons (*Antirrhinum*) 'Avalanche' (optional)
Ivy-leaved geranium (*Pelargonium*) 'Blue Beard'
Ivy-leaved geranium (*Pelargonium*) 'L'Elégante' (optional)
3 pink petunias

snapdragon

ivy-leaved geraniums

pink petunia

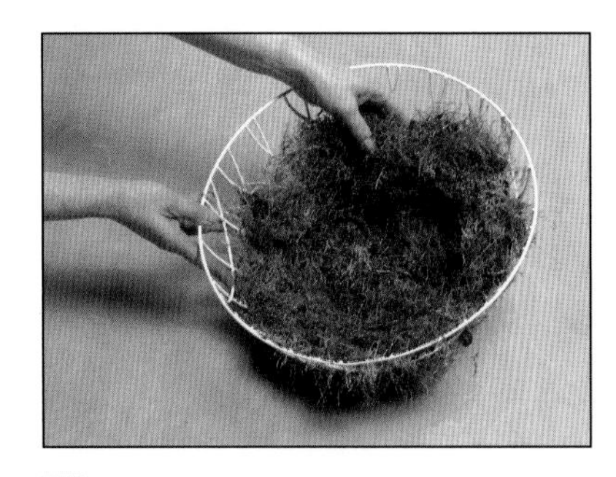

1 Line the lower half of the basket with moss.

2 Plant the snapdragons in the side of the basket, resting the rootballs on the moss and guiding the foliage through the side of the basket.

3 Line the remainder of the basket with moss, tucking it carefully around the underplanted snapdragons.

4 Fill the basket with compost, mixing a teaspoon of slow-release plant food granules into the top layer of compost. Plant the 'Blue Beard' geranium (*Pelargonium*) at the back of the basket.

5 Plant the 'L'Elégante' geranium at the front of the basket.

GARDENER'S TIP

It is a good idea to include a number of different plants in a hanging basket. It creates a more interesting picture and ensures that if one plant does not thrive, as happened to the snapdragons in this basket, the other plants will still make a good display.

Plant in late spring or early summer

6 Plant the petunias around the geraniums. Water thoroughly and hang in a sunny position.

A Pastel Combination

A white-flowered geranium is planted with silver *Senecio*, white *Bacopa* and pink *Diascia* to make a delicate planting scheme for this basket. This type of basket works well against a dark background.

MATERIALS
30 cm (12 in) hanging basket
Sphagnum moss
Compost
Slow-release plant food granules

PLANTS
3 *Bacopa* 'Snowflake'
3 pink *Diascia*
3 *Senecio cineraria* 'Silver Dust'
White-flowered ivy-leaved geranium (*Pelargonium*)

1 Line three-quarters of the basket with moss.

2 Partially fill the lined area with compost and plant one of the *Bacopa* into the side of the basket. Rest the rootball on the compost and gently feed the foliage through the side.

3 Plant one of the *Diascia* into the side of the basket in the same way.

4 To complete the underplanting, plant a *Senecio* into the side of the basket.

5 Line the rest of the basket with moss and top up with compost, mixing a teaspoon of slow-release plant food granules into the top of the compost. Firm well to ensure the plants in the side of the basket are securely in place. Plant the geranium (*Pelargonium*) in the centre.

6 Fill in around the geranium with the remaining plants. Water thoroughly and hang in a sunny position.

GARDENER'S TIP

When the summer is over you can save the geranium for next year by digging it up, cutting it back to about 15 cm (6 in) and potting it up. It can be over-wintered indoors or in a greenhouse. Keep fairly dry.

Plant in late spring or early summer

Flowers for Late Summer

Although this window box is already looking good, towards the end of the summer it will really come into its own – by then the vibrant red and purple flowers of the geranium, *Salvia* and lavenders will be at their most prolific.

MATERIALS
60 cm (24 in) wooden planter, stained black
Polystyrene or other suitable drainage material
Compost
Slow-release plant food granules

PLANTS
Geranium (*Pelargonium*) 'Tomcat'
2 *Lavandula pinnata*
2 *Salvia* 'Raspberry Royal'
2 blue *Brachycome* daisies
Convolvulus sabatius
6 rose-pink alyssum

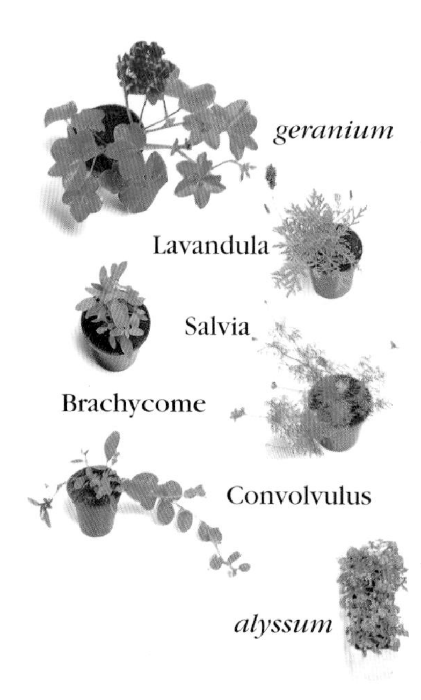

1 Line the base of the container with polystyrene or similar drainage material. Fill the window box with compost, mixing in 3 teaspoons of slow-release plant food granules. Plant the geranium at the back of the window box, in the centre.

2 Plant the two lavenders in the rear corners of the box.

3 Plant the *Salvia* at the front of the box on either side of the geranium.

4 Plant the *Brachycome* daisies in the front corners of the box.

5 Plant the *Convolvulus* in the centre, in front of the geranium.

GARDENER'S TIP

Both the lavenders and the *Salvia* are highly aromatic, so if possible position this box near a door or a path, so that you can enjoy the fragrance as you brush against the plants.

Plant in early summer

6 Fill in the spaces with the alyssum. Water well and place in a sunny position.

Begonias and Fuchsias

Fuchsias are wonderful hanging basket plants as they flower prolifically late into the autumn. By the end of summer, when the other plants may start to look a bit straggly, the fuchsia will be at its best with a glorious display of colour.

MATERIALS
36 cm (14 in) hanging basket
Sphagnum moss
Compost
Slow-release plant food granules

PLANTS
2 *Diascia* 'Ruby Field'
3 *Helichrysum microphyllum*
Fuchsia 'Rose Winston' or similar soft pink
3 deep pink begonia

begonia

Helichrysum

Diascia

Fuchsia

begonia

GARDENER'S TIP

If some of the plants in the basket begin to look straggly in comparison with the fuchsia, cut them right back and give a liquid feed – they will grow with renewed vigour and provide a wonderful autumn show.

Plant in late spring or early summer

1 Line the lower half of the basket with moss and arrange the *Diascia* and *Helichrysum* in the basket to decide where to plant each one before entangling them in the wires.

4 Fill the basket with compost. Mix a teaspoon of slow-release plant food granules into the top of the compost. Plant the fuchsia in the centre of the basket.

2 Plant the two *Diascia* into the sides of the basket by resting the rootballs on the moss and gently feeding the foliage through the wire.

3 Line the rest of the basket with moss, partly fill with compost and plant the three *Helichrysum* into the side of the basket near the rim using the same method as before.

5 Finally, plant the three begonias around the fuchsia. Water well and hang the basket in full sun or partial shade.

Pot of Sunflowers

Sunflowers grow very well in pots provided you are not growing the giant varieties. Grow your own from seed; there are many kinds to choose from, including the double flowers used here.

MATERIALS AND TOOLS
Large glazed pot, 30 cm (12 in) diameter
Polystyrene or similar drainage material
Equal mix loam-based compost and container compost
Slow-release plant food granules

PLANTS
3 strong sunflower seedlings, approximately 20 cm (8 in) tall

sunflower seedling

1 Line the base of the pot with drainage material.

2 Fill the pot with the compost mix, pressing down so that there are no air spaces. Scoop out evenly spaced holes for each seedling and plant, firming the compost around the plants.

3 Scatter 1 tablespoon of plant food granules on the surface of the compost. Place in a sunny position, out of the wind, and water regularly.

GARDENER'S TIP

Allow at least one of the sunflower heads to set seed. As the plant starts to die back, cut off the seedhead and hang it upside-down to ripen. Reserve some seeds for next year and then hang the seedhead outside for the birds.

Plant seeds in spring and small seedlings in summer to flower in late summer.

Heather Window Box

This is a perfect project for an absolute beginner as it is extremely simple to achieve. The bark window box is a sympathetic container for the heathers, which look very much at home in their bed of moss.

MATERIALS AND TOOLS
Bark window box, 30 cm (12 in) long
Crocks or similar drainage material
Ericaceous compost
Bun moss

PLANTS
3 heathers

GARDENER'S TIP

Do not be tempted to use ordinary compost as it contains lime, which, with a very few exceptions, is not suitable for the majority of heathers.

Plant in autumn.

1 Put a layer of crocks or similar drainage material in the bottom of the box.

2 Remove the heathers from their pots and position them in the window box.

3 Fill the gaps between the plants with the compost, pressing it around the plants. Water.

4 Tuck the bun moss snugly around the plants so that no soil is visible. Place in sun or partial sun.

Evergreens and Extra Colour

They may be easy to look after but all-year-round window boxes can start to look a bit lifeless after a couple of seasons. It does not take much trouble to add a few seasonal flowers and it can make all the difference.

GARDENER'S TIP

At the end of the summer, remove the *Diascias* and marguerite, feed the remaining plants with more granules and fill the spaces with winter-flowering plants such as pansies or heathers.

Plant in spring

MATERIALS

76 cm (30 in) plastic window box
Compost
Slow-release plant food granules

PLANTS

Hebe 'Baby Marie'
Convolvulus cneorum
Potentilla 'Nunk'
Variegated ivies
2 *Diascia* 'Ruby Field'
Pink marguerite (*Argyranthemum*) 'Flamingo'

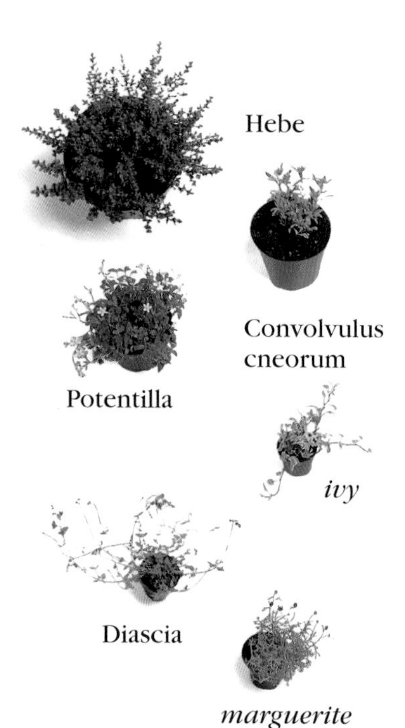

1 Check the drainage holes are open in the base and, if not, drill or punch them open. Fill the window box with compost, mixing in 3 teaspoons of slow-release plant food granules. Plant the *Hebe* in the centre.

2 Plant the *Convolvulus* near the right-hand end of the window box.

3 Plant the *Potentilla* near the left-hand end of the window box.

4 Plant the two ivies at the front corners of the window box.

5 Plant the *Diascias* on either side of the *Hebe* at the front of the window box.

6 Plant the marguerite between the *Hebe* and the *Convolvulus* at the back of the window box. Water well and stand in full or partial sun.

Autumn Colour

Long Tom pots were very popular in Victorian times. They are now being made again and are available in a variety of colours. There is a wonderfully architectural quality to the outline of this cream Long Tom and the plants it contains.

Gardener's Tip

These plants will look marvellous for one season, but after that they will benefit from being repotted into a larger container.

Plant in spring or early summer to flower in autumn.

MATERIALS AND TOOLS
Long Tom, 30 cm (12 in) diameter
Crocks
Equal mix loam-based compost and standard compost
Clay saucer
Gravel
Slow-release plant food granules
Trowel or scoop

PLANTS
Cape figwort (*Phygelius*)
Day lily (*Hemerocallis dumortieri*)
Hardy fuchsia, preferably a variety with compact growth

day lily

Cape figwort

fuchsia

1 Cover the drainage holes at the base of the pot with crocks.

2 Position the largest plant first. You will probably have to loosen the soil around the rootballs of the plants to fit them all in the pot. Gently squeeze the soil and tease the roots loose.

3 Add the other two plants, again loosening the soil if necessary.

4 Fill any spaces between the plants with compost, and push the compost firmly down the sides of the pot so that no air spaces are left.

5 The shape of the container means that the soil will dry out quite quickly, especially when so densely planted. To counteract this, stand it in a clay saucer filled with wet gravel.

6 Scatter 1 tablespoon of plant food granules onto the compost and mix in. Water regularly and place in a sunny position.

Alpine Sink

An old stone sink is a perfect container for a collection of Alpine plants. The rock helps create the effect of a miniature landscape and provides shelter for some of the plants. The sink is set up on the stand of an old sewing machine so that the beauty of the tiny plants can be admired easily.

MATERIALS AND TOOLS
A stone sink or trough,
75 cm × 50 cm (30 in × 20 in)
Crocks
Moss-covered rock
Loam-based compost with ⅓ added coarse grit
Washed gravel
Trowel

PLANTS
Achillea tomentosa
Veronica penduncularis
Ivy
Sedum spathulifolium purpureum
Hebe
Sedum ewersii
Aster natalensis
Alpine willow (*Salix alpina*)
Arabis ferdinandi-coburgii 'Variegata'

1 Cover the drainage hole of the sink with crocks.

2 Position the rock. It is important to do this before adding the soil to create the effect of a natural rocky outcrop. This will never be achieved if you place your rock onto the surface of the compost.

3 Pour the compost into the sink. (If you are lucky enough to have a moss-covered sink, like the one used here, be careful not to disturb the moss or cover it with compost.)

Sedum ewersii

Achillea tomentosa

Sedum spathulifolium purpureum

Hebe

Veronica penduncularis

ivy

4 Plan the positions of your plants so that the end result will have a good balance of shape and colour. Start planting from one end. As the sink is very shallow you will need to scoop out the soil right to the base before planting. Alpine plants are used to shallow soil so this will not cause any problems.

GARDENER'S TIP

If you do not have a stone sink you could use a butler's sink instead. You could even give it a stone effect by covering it with a mix of Portland cement, sand and peat in proportions of 1 to 2.5 to 1.5. You will need to cover the whole surface with impact bonding PVA glue before applying the mix to the sink. Build up the layers gradually, wearing rubber gloves.

Plant in spring to flower in autumn.

5 Be careful to ensure that the bottom leaves of low-growing plants are level with the soil. Too low and they will rot; too high and they will dry out.

6 When all the plants are in place, carefully pour washed gravel all around them, lifting leaves to cover the whole soil area. Water and place in full or partial sun.

Classic Winter Colours

Convolvulus cneorum **is an attractive small shrub with eye-catching silver-grey leaves which last through winter and it has white flowers in spring and summer. Planted with ice-blue pansies, it makes a softly subtle display from autumn to spring.**

MATERIALS
30 cm (12 in) hanging basket
Sphagnum moss
Compost

PLANTS
8 silver/blue pansies (*Viola*) 'Silver Wings', or similar
Convolvulus cneorum

Convolvulus

pansies

GARDENER'S TIP

At the end of winter cut back any dead wood or straggly branches on the *Convolvulus cneorum* and give a liquid feed to encourage new growth. Small shrubs such as this may be used in hanging baskets for one season, but will then need planting into a larger container or the border.

Plant in autumn

1 Half line the basket with moss and fill with compost to the top of the moss.

2 Plant four of the pansies into the side of the basket by placing their root balls on the compost and gently guiding the leaves through the side of the basket.

3 Line the rest of the basket with moss and top up with compost. Plant the *Convolvulus* in the centre of the basket.

4 Plant the remaining four pansies around the *Convolvulus*. Water well and hang in sun or partial shade.

Evergreen Garden

Evergreen plants come in many shapes, sizes and shades of green. Grouped together in containers they will provide you with year-round interest and colour.

MATERIALS AND TOOLS
Terracotta containers of various sizes
Crocks or similar drainage material
Equal mix loam-based compost and container compost
Saucers
Gravel
Slow-release plant food granules
Trowel

PLANTS
False cypress (*Chamaecyparis*)
Silver *Euonymus*
Darwin's barberry (*Berberis darwinnii*)
Barberry (*Berberis atropurpurea nana*)
Cypress (*Cupressus filifera aurea*)
Pachysandra terminalis
Bergenia

1 Large plants, such as *Chamaecyparis*, should be potted into a proportionally large container. If it is at all potbound, tease the roots loose before planting in its new pot. Place plenty of crocks or similar drainage material at the base of the pot. Fill around the rootball with compost, pressing it down firmly around the edges of the pot.

2 Smaller plants, like *Bergenia*, should be planted in a pot slightly larger than its existing pot. Place crocks in the base of the pot, position the plant and then fill around the edges with compost. Repeat with the remaining plants.

3 Plants will stay moist longer if they are stood in saucers of wet gravel. This group of plants will do well positioned in partial shade. Water regularly and feed with slow-release plant food granules in the spring and autumn.

GARDENER'S TIP

Include some golden or variegated foliage amongst your evergreens or the group will look rather dull and one dimensional. Experiment for yourself and see how the lighter colours "lift" a group of plants.

Plant at any time of the year.

Gothic Ivy

Twisted willow branches set into a chimney pot offer an attractive support for ivy and will provide welcome interest in the winter.

MATERIALS AND TOOLS
Chimney pot
Standard compost
1 m (1 yard) wire netting

PLANTS
4 or 5 branches of twisted willow
Large ivy (*Hedera helix* var. *hibernica* was used here)

ivy

GARDENER'S TIP

You may find that some of your twisted willow branches take root in the compost. Plant a rooted branch in the garden where it will grow into a tree. It will eventually be quite large so don't plant it near the house.

Plant at any time of the year.

1 Place the chimney pot in its final position (in shade or half-shade) and half fill with compost. Fold or crumple the wire netting and push down into the chimney pot so that it rests on the compost.

2 Arrange the willow branches in the chimney pot, pushing the stems through the wire netting so that they stay in place.

3 Rest the ivy, in its pot, on the wire netting amongst the willow branches. Fill the chimney pot with compost to within 10 cm (4 in) of the rim.

4 Carefully cut loose any ties and remove the supporting cane.

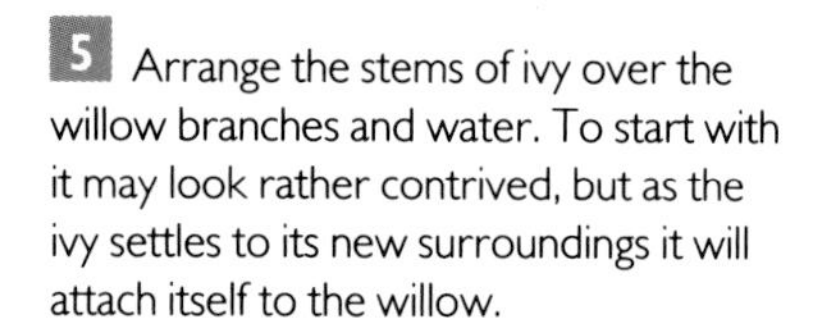

5 Arrange the stems of ivy over the willow branches and water. To start with it may look rather contrived, but as the ivy settles to its new surroundings it will attach itself to the willow.

Year-round Window Box

In the same way that a garden has certain plants that provide structure throughout the year, this window box has been planted so that there is always plenty of foliage. Extra colour may be introduced each season by including small flowering plants, such as heathers.

MATERIALS AND TOOLS
Window box, 1 m (1 yard) long, preferably self-watering
Equal mix loam-based compost and container compost
Slow-release plant food granules
Bark mulch
Trowel

PLANTS
Skimmia rubella
2 *Arundinaria pygmaea*
2 *Cotoneaster conspicuus*
2 periwinkle (*Vinca variegata*)
6 heathers

Arundinaria pygmaea

Skimmia rubella

Cotoneaster conspicuus

heather

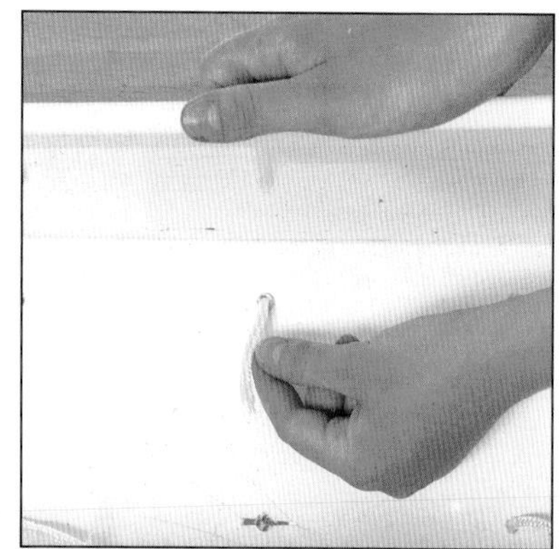

1 Feed the wicks through the base of the plastic liner.

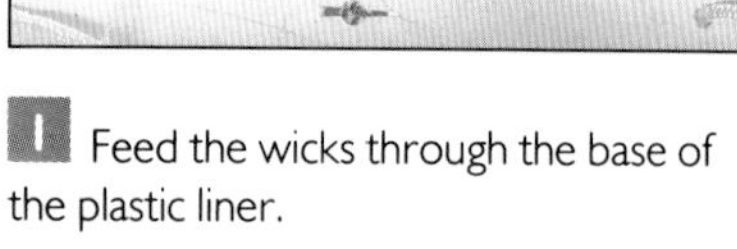

Note Steps 1 and 2 are for self-watering containers only.

periwinkle

2 Slip the liner into the wooden window box.

3 Before you start planting, plan the positions of the plants so that the colours and shapes look well balanced.

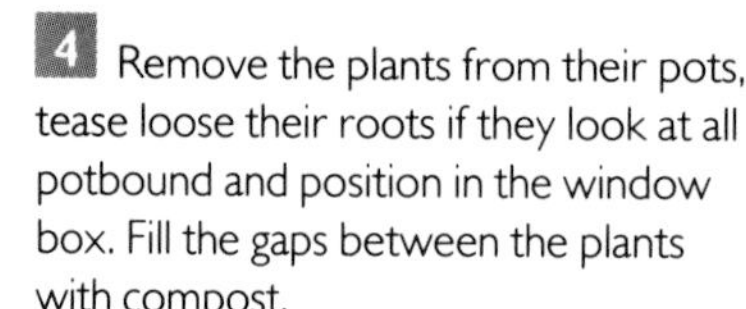

4 Remove the plants from their pots, tease loose their roots if they look at all potbound and position in the window box. Fill the gaps between the plants with compost.

5 Once the structure plants are in place you can add the colour, in this case the heathers. Scoop out a hole for each heather and then plant, pressing firmly around each one. Scatter two tablespoons of plant food granules along the surface.

6 Top-dress the window box with a layer of bark; this will help conserve moisture and prevent soil splashes on the leaves. Water.

GARDENER'S TIP

Plants do not need watering in winter, unless they are in a position where the rain does not reach the container. Even then they should be watered sparingly and not in frosty weather. Self-watering containers should be drained before winter to prevent frost damage.

Plant at any time of the year.

Classic Topiary

The clean lines of the topiary are matched by the simplicity of the terracotta pots. The eye is drawn to the outlines of the box plants so decorated pots would be an unnecessary distraction.

MATERIALS AND TOOLS
4 large terracotta pots
Bark mulch
Crocks
Equal mix loam-based compost and standard compost
Slow-release plant food granules
Trowel

PLANTS
4 box trees (*Buxus*) in different topiary shapes

three-ball topiary *ball topiary*

corkscrew topiary

GARDENER'S TIP

Don't get carried away when you trim topiary. Little and often with an ordinary pair of scissors is better than occasional dramatic gestures with a pair of shears.

Plant at any time of the year.

1 If the plant has been well looked after in the nursery it may not need potting on yet. In this case simply slip the plant in its pot into the terracotta container.

2 To conserve moisture and conceal the plastic pot, cover with a generous layer of bark.

3 To repot a box tree, first place a good layer of crocks in the bottom of the pot.

4 Remove the tree from its plastic pot and place it in the terracotta pot. Surround the rootball with compost.

5 Push the compost down the side of the pot to ensure that there are no air spaces.

6 Scatter a tablespoon of plant food granules on the surface of the pot and then top with a good layer of bark. Water well and regularly. Position in sun or partial shade.

Winter Cheer

Many window boxes are left unplanted through the winter, but you can soon brighten the house or garden for the winter season with this easy arrangement of pot-grown plants plunged in bark.

MATERIALS
40 cm (16 in) glazed window box
Bark

PLANTS
2 miniature conifers
2 variegated ivies
2 red polyanthus

1 Water all the plants. Place the conifers, still in their pots, at either end of the window box.

2 Half-fill the window box with bark.

3 Place the pots of polyanthus on the bark between the two conifers.

4 Place the pots of ivy on the bark in the front corners of the window box. Add further bark to the container until all the pots are concealed. Water only when plants show signs of dryness. Stand in any position.

GARDENER'S TIP

When it is time to replant the window box, plunge the conifers, still in their pots, in a shady position in the garden. Water well through the spring and summer and they may be used again next year.

Plant in early winter

Golden Christmas Holly

Evergreen standard holly trees are splendid container plants. This golden holly has been dressed up for Christmas with bows and baubles in a gilded pot.

Materials and Tools
Terracotta pot, 40 cm (16 in) high
Gold spray-paint
Crocks or similar drainage material
Composted manure
Loam-based compost
Pine cones
1 m (1 yard) wired gold-mesh ribbon, for bow
Selection of tin Christmas decorations, sprayed gold

Plants
Golden holly

golden holly

1 Spray the pot with gold paint and leave to dry.

2 Place a good layer of crocks or other drainage material in the base of the pot.

3 Cover with an 8 cm (3 in) layer of composted manure and a thin layer of loam-based potting compost. Remove the holly from its existing container and place in the gilded pot, surround the rootball with compost, pressing down firmly to ensure that the tree is firmly planted.

4 Surround the base of the tree with pine cones.

5 Tie a bow with the ribbon and attach it to the trunk of the tree. Hang the decorations in the branches. Water the tree to settle it in, but don't do this on a frosty day or the water will freeze.

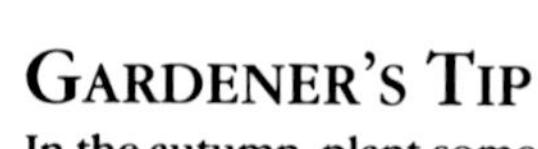

Gardener's Tip

In the autumn, plant some corms of Reticulata irises or similar small bulbs in the compost surrounding the tree for a delightful spring display.

Plant in autumn, winter or spring.

An Evergreen Wall Basket

Pansies will flower throughout the winter. Even if they are flattened by rain, frost or snow, at the first sign of improvement in the weather their heads will pop up again to bring brightness to the dullest day. They have been planted with ivies to provide colour from early autumn through to late spring.

MATERIALS
30 cm (12 in) wall basket
Sphagnum moss
Compost

PLANTS
2 golden variegated ivies
2 copper pansies (*Viola*)
Yellow pansy (*Viola*)

pansy

ivy

GARDENER'S TIP

Winter baskets do not need regular feeding and should only be watered in very dry conditions. To prolong the flowering life of the pansies, dead-head regularly and pinch out any straggly stems to encourage new shoots from the base.

Plant in autumn

1 Line the basket with moss.

2 Three-quarters fill the basket with compost and position the ivies with their rootballs resting on the compost. Guide the stems through the sides of the basket so that they trail downwards. Pack more moss around the ivies and top up the basket with compost.

3 Plant the two copper pansies at either end of the basket.

4 Plant the yellow pansy in the centre of the basket. Water well and hang in shade or semi-shade.

Trug of Winter Pansies

Winter pansies are wonderfully resilient and will bloom bravely throughout the winter as long as they are regularly deadheaded. This trug may be moved around to provide colour wherever it is needed and acts as a perfect antidote to mid-winter gloom.

MATERIALS AND TOOLS
Old wooden trug
Sphagnum moss
Standard compost
Slow-release plant food granules
Trowel or plastic pot

PLANTS
15 winter-flowering pansies

winter-flowering pansies

GARDENER'S TIP

Not everyone has an old trug available, but an old basket, colander, or an enamel bread bin could be used instead. Junk shops and flea markets are a great source of containers that are too battered for their original use, but great for planting.

Plant in autumn to flower in winter.

1 Line the trug with a generous layer of sphagnum moss.

2 Fill the moss lining with compost.

3 Plant the pansies by starting at one end and filling the spaces between the plants with compost as you go. Gently firm each plant into position and add a final layer of compost mixed with a tablespoon of plant food granules around the pansies. Water and place in a fairly sunny position.

Cascading Ivy with Dark Flowers and Foliage

Viola 'Bowles' Black' is combined with black grass – *Ophiopogon* – and the dramatically coloured *Begonia rex*. Pale pink *Nemesia* and variegated ground ivy provide an effective contrast.

MATERIALS
36 cm (14 in) hanging basket
Sphagnum moss
Compost
Slow-release plant food granules

PLANTS
Begonia rex
black grass (*Ophiopogon*)
2 variegated ground ivies (*Glechoma hederacea* 'Variegata')
2 *Nemesia* 'Confetti'
2 *Viola* 'Bowles' Black'

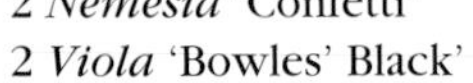

1 Line the basket with moss.

2 Fill the basket with compost, mixing a teaspoon of plant food granules into the top layer of compost. Plant the begonia at the back of the basket.

3 Plant the black grass in front of the begonia.

4 Plant the ground ivies at either side of the basket, angling the rootballs to encourage the foliage to tumble down the sides of the basket.

5 Plant the *Nemesia* on either side of the begonia.

GARDENER'S TIP

At the end of the season the begonia can be potted up and kept indoors as a houseplant and the black grass can be planted in an outdoor container.

Plant in late spring or early summer

6 Plant the violas on either side of the black grass. Water well and hang in light shade.

A Topiary Planting

Topiary box plants remain in their pots in this window box. A mulch of bark conceals the pots and retains moisture, and small pots of white *Bacopa* add another dimension to this simple design.

MATERIALS
64 cm (25 in) terracotta planter
Bark

PLANTS
Box pyramid in 5-litre (9 in) pot
2 box balls in 5-litre (9 in) pots
5 pots white *Bacopa*

1 Water all the plants thoroughly. Stand the box pyramid in the centre of the container.

2 Stand the box balls on either side of the pyramid.

3 Fill the container with bark.

4 Plunge the pots of *Bacopa* in the bark at the front of the container. Stand in light shade. Water regularly.

GARDENER'S TIP

Provided the box plants are not root-bound they will be quite happy in their pots for a year. If the leaves start to lose their glossy dark green colour, it is a sign that the plants need a feed. Sprinkle a long-term plant feed on the surface of the pots and boost with a liquid feed.

Plant box at any time of year, and Bacopa *in spring*

Seasonal Planting Lists for Containers

Plants for an early spring window box
Crocuses
Ferns
Forget-me-nots (*Myosotis*)
Hyacinths
Ivies
Narcissi and daffodils
Pansies
Periwinkle (*Vinca minor*)
Polyanthus
Primroses
Tulips
Violas

Plants to last into late autumn
Begonias
Felicia
Fuchsias
Geraniums (*Pelargonium*)
Helichrysum
Salvia
Scaevola

Plants for the summer
Alyssum
Brachycome daisies
Busy lizzies
Campanula
Convolvulus
Dianthus
Felicia
Fuchsias
Heliotropes
Lavenders
Lobelia
Marguerites
Nasturtiums
Osteospermum
Petunias
Pot marigolds (*Calendula*)
Salvia
Tobacco, flowering (*Nicotiana*)
Verbenas
Violas

Plants for winter window boxes
Convolvulus
Ivies
Miniature conifers
Ornamental cabbages
Pansies
Periwinkle (*Vinca minor*)
Polyanthus
Violas

Seasonal Planting Lists for Baskets

Plants for an early spring basket
Crocuses
Miniature daffodils and narcissus
Ferns
Forget-me-nots (*Myosotis*)
Ivies
Pansies
Periwinkle (*Vinca minor*)
Polyanthus
Primroses
Violas

Plants for the summer
Ageratum
Alyssum
Begonia semperflorens
Brachycome
Calendula (dwarf)
Campanula isophylla
Convolvulus sabatius
Dianthus
Diascia
Erigeron
Felicia
Fragaria
Fuchsias
Gazanias
Geranium *(Pelargonium)*
Hedera
Heliotrope
Lantana
Lavandula (dwarf)
Lobelia
Nasturtiums
Pansies
Parsley
Periwinkle (*Vinca min*or)
Petunias
Salvia
Scabiosa
Scaevola
Thymus
Verbena
Viola

Plants to last into late autumn
Begonias
Felicia
Fuchsias
Geraniums (*Pelargonium*)
Helichrysum
Salvia
Scaevola

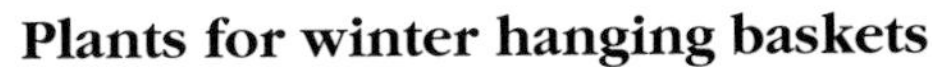

Plants for winter hanging baskets
Convolvulus cneorum
Ivies
Pansies
Periwinkle (*Vinca minor*)
Violas

INDEX